Point group character tables and related data

J. A. SALTHOUSE and M. J. WARE

Lecturers in Chemistry, University of Manchester

Cambridge · At the University Press, 1972

Published by the Syndics of the Cambridge Univ Press
Bentley House, 200 Euston Road, London NW1 2DB
American Branch: 32 East 57th Street, New York, N.Y. 10022

Library of Congress Catalogue Card Number: 72-75774
ISBN: 0 521 08139 4

Printed in Great Britain by
W. Heffer & Sons Limited,
Cambridge, England.

CONTENTS

Foreword

The growing application of spectroscopy to chemical research over the last ten years has resulted in a wider use of group theory, and the presentation of this topic is gaining acceptance among teachers of chemistry. This changing attitude is reflected in the recent appearance of several good introductory textbooks. Our experience of teaching group theory to both undergraduate and postgraduate chemistry students at Manchester University has shown that a lecture course, complemented by the use of a suitable text, rapidly enables the student to gain a competence in the subject sufficient for his needs. The value of the elementary textbook then lies mainly in its appendices of essential tables. Therefore it seems desirable to publish these data separately in an inexpensive form, on the presumption that a user, whether student or research worker, will be familiar with the theoretical background and method of application.

This little book can lay no claim to originality; indeed, we must record our indebtedness to the many authoritative works on which we have drawn heavily. Regrettably these are too numerous for individual acknowledgement here, but we would urge the reader to compensate himself for our shortcomings by consulting the texts listed in the general bibliography.

Within the bounds of reasonable chemical possibility, we have tried to make the point group data complete. Most earlier authors have justified omissions by arguing the improbability of any chemical species belonging to the absent point groups. There seems lately to have grown up a sport among preparative chemists to synthesise just those molecules whose point group character tables languish in obscurity. Accordingly, we have attempted to anticipate their activity by including explicit character tables for all the cubic and icosahedral groups and for axial groups of seven- and eight-fold symmetry. The possibility that the hyper-Raman effect may emerge as an important spectroscopic technique has persuaded us to include the transformations of the first hyperpolarisability tensor; these also serve to indicate the symmetry species of the f-orbitals.

On the other hand we have not attempted to cater for the theoretician who requires an adequate treatment of the pure rotation group and detailed information on the properties of angular momentum.

In conclusion, we hope that this compilation will prove a convenient source of character tables for the student and a useful *vade-mecum* for the practising spectroscopist.

December 1971

J. A. Salthouse
M. J. Ware

1 Numerical data

1.1 Physical constants

The following list draws mainly from the self-consistent set of values[1] recommended in 1969, but also includes the SI equivalents of certain quantities, marked with an asterisk, whose use as units is either discontinued or restricted to specialised contexts.[2] Estimated uncertainties in parentheses apply to the last digit(s) of the preceding number and represent three standard deviations.

Symbol	Quantity	Value in SI units	Value in other units
c	speed of light in vacuo	2.997 925 0 (30)×10^{8} m s^{-1}	2.997 925 0 (30)×10^{10} cm s^{-1}
e	elementary charge	1.602 192 (21)×10^{-19} C	4.803 250 (63)×10^{-10} esu
k	Boltzmann constant	1.380 62 (18)×10^{-23} J K^{-1}	1.380 62 (18)×10^{-16} erg deg^{-1}
h	Planck constant	6.626 20 (15)×10^{-34} J s	6.626 20 (15)×10^{-27} erg s
$\hbar$	$(\frac{h}{2\pi})$	1.054 592 (24)×10^{-34} J s	1.054 592 (24)×10^{-27} erg s
N_A	Avogadro constant	6.022 17 (12)×10^{23} mol^{-1}	
R	gas constant $(N_A k)$	8.314 3 (12) J mol^{-1}K^{-1}	1.987 2 cal deg^{-1}mol^{-1}
F	Faraday constant $(N_A e)$	9.648 67 (16)×10^{4} C mol^{-1}	2.892 599 (48)×10^{14} esu mol^{-1}
T_{ice}	Ice point temperature	273.150 0 (1) K	
RT_{ice}		2.271 06 (12)×10^{3} J mol^{-1}	5.428 04×10^{2} cal mol^{-1}
atm	*standard atmospheric pressure	1.013 25×10^{5} N m^{-2}	1.013 25×10^{6} dyn cm^{-2}
V_0	molar volume of perfect gas	2.241 36 (30)×10^{-2} m^{3}mol^{-1}	22.413 6 dm^{3}mol^{-1}
μ_0	permeability of free space	4π×10^{-7} J s^{2}C^{-2}m^{-1} (H m^{-1})	
ε_0	permittivity of free space $(1/\mu_0 c^2)$	8.854 185 (18)×10^{-12} J^{-1}C^{2}m^{-1}	(F m^{-1})
m_p	rest mass of proton	1.672 614 (33)×10^{-27} kg	1.007 276 61 (24) u
m_n	rest mass of neutron	1.674 920 (33)×10^{-27} kg	1.008 665 20 (10) u
m_e	rest mass of electron	9.109 56 (11)×10^{-31} kg	5.485 93 (10)×10^{-4} u
a_0	Bohr radius $(h^2/\pi\mu_0 c^2 m_e e^2)$	5.291 772 (24)×10^{-11} m	0.529 177 2 (24) Å
r_e	electron radius $(\mu_0 e^2/4\pi m_e)$	2.817 939 (39)×10^{-15} m	
R_∞	Rydberg constant $(\mu_0^2 m_e e^4 c^3/8h^3)$	1.097 373 12 (33)×10^{7} m^{-1}	1.097 373 12 (33)×10^{5} cm^{-1}
R_H	$(R_\infty/[1+m_e/m_p])$	1.096 775 8 (3)×10^{7} m^{-1}	1.096 775 8 (3)×10^{5} cm^{-1}
μ_B	Bohr magneton $(eh/4\pi m_e)$	9.274 10 (20)×10^{-24} A m^{2} (J T^{-1})	9.274 10 (20)×10^{-21} erg gauss^{-1}
μ_N	nuclear magneton $(\mu_B m_e/m_p)$	5.050 95 (15)×10^{-27} A m^{2} (J T^{-1})	5.050 95 (15)×10^{-24} erg gauss^{-1}
γ	gyromagnetic ratio of proton	2.675 197 (25)×10^{8} rad s^{-1}T^{-1}	2.675 197 (25)×10^{4} rad s^{-1}gauss^{-1}
σ	Stefan constant $(2\pi^5 k^4/15h^3c^2)$	5.669 7 (29)×10^{-8} W m^{-2}K^{-4}	5.669 7 (29)×10^{-5} erg cm^{-2}s^{-1}deg^{-4}
α	fine structure constant $(\mu_0 e^2 c/2h)$	7.297 351 (33)×10^{-3}	
G	gravitational constant	6.673 2 (93)×10^{-11} N m^{2}kg^{-2}	6.673 2 (93)×10^{-8} cm^{3}g^{-1}s^{-2}
g_n	standard gravitational acceleration	9.806 65 m s^{-2}	980.665 cm s^{-2}
eV	*electronvolt	1.602 192 (21)×10^{-19} J	1.602 192 (21)×10^{-12} erg
D	*Debye unit	3.335 6×10^{-30} C m	10^{-18} esu cm

Symbol	Quantity	Value in SI units	Value in other units
cal_{th}	*calorie (thermochemical)	4.184 00 J	$4.184\,00\times10^{7}$ erg
cal_{IT}	*calorie (international steam table)	4.186 8 J	$4.186\,8\times10^{7}$ erg
$cal_{15°}$	*calorie (15° Centigrade)	4.185 5 J	$4.185\,5\times10^{7}$ erg
mmHg	*millimetre of mercury ($13.595\,1 g_n$)	$133.322\,39$ N m^{-2}	$1.333\,223\,9\times10^{3}$ dyn cm^{-2}
Torr	*torr (atm/760)	$133.322\,37$ N m^{-2}	$1.333\,223\,7\times10^{3}$ dyn cm^{-2}
u	*unified atomic mass unit	$1.660\,531(33)\times10^{-27}$ kg	$1.660\,531(33)\times10^{-24}$ g

Numerical constants

π	3.141 592 653 590	π^2	9.869 604 401 089	$\frac{1}{\pi}$	0.318 309 886 184
e	2.718 281 828 459	$\log_e 10$	2.302 585 092 994	$\log_{10} e$	0.434 294 481 903

Conversion factors

1 radian = 57.295 779 513 082 degrees 1 degree = 0.017 453 292 519 943 radians

$4\pi^2c^2u = 5.891\,80 \times 10^{-9}$ N m $= 0.589\,180 \times 10^{-6}$ N cm (1 N cm^{-1} = 1 millidyn $Å^{-1}$)

$hc/kT_{ice} = 5.267\,57 \times 10^{-3}$ cm $hc/kT_{25°C} = 4.825\,88 \times 10^{-3}$ cm

$h/8\pi^2c = 2.799\,33 \times 10^{-44}$ kg m $= 2.799\,33 \times 10^{-39}$ g cm $h/8\pi^2cu = 1.685\,80 \times 10^{-15}$ cm

$h/8\pi^2u = 5.053\,91 \times 10^{-9}$ m^2s^{-1} $= 5.053\,91 \times 10^{9}$ pm^2MHz

1.2 Energy conversion table

In the column headings of the following table W denotes energy, and the other physical quantities and their units are represented by the recommended symbols.[2] The conversion factors consist of a numerical part followed, if necessary, by the power of 10 which multiplies this, e.g. 6.022 52 E+23 = $6.022\,52 \times 10^{23}$.

Physical quantity / Unit	W/J	($W_m = N_A W$) W_m/J mol^{-1}	($\tilde{\nu}$ = W/hc) $\tilde{\nu}$/cm^{-1}	W/eV	W/cal$_{th}$	($W_m = N_A W$) W_m/cal$_{th}$ mol^{-1}	(T = W/k) T/K	(ν = W/h) ν/MHz	(H = W/μ_B) H/G
1 J	1	6.022 17 E+23	5.034 02 E+22	6.241 45 E+18	2.390 06 E-1	1.439 33 E+23	7.243 12 E+22	1.509 16 E+27	1.078 27 E+27
1 J mol^{-1}	1.660 53 E-24	1	8.359 14 E-2	1.036 41 E-5	3.968 76 E-25	2.390 06 E-1	1.202 74 E-1	2.506 01 E+3	1.790 50 E+3
1 cm^{-1}	1.986 48 E-23	1.196 29 E+1	1	1.239 85 E-4	4.747 80 E-24	2.859 21	1.438 83	2.997 92 E+4	2.141 97 E+4
1 eV	1.602 19 E-19	9.648 69 E+4	8.065 49 E+3	1	3.829 33 E-20	2.306 09 E+4	1.160 49 E+4	2.417 96 E+8	1.727 60 E+8
1 cal$_{th}$	4.184 00	2.519 68 E+24	2.106 24 E+23	2.611 42 E+19	1	6.022 17 E+23	3.030 52 E+23	6.314 33 E+27	4.511 49 E+27
1 cal$_{th}$ mol^{-1}	6.947 67 E-24	4.184 00	3.497 47 E-1	4.336 34 E-5	1.660 53 E-24	1	5.032 28 E-1	1.048 51 E+4	7.491 48 E+3
1 K	1.380 62 E-23	8.314 33	6.950 09 E-1	8.617 07 E-5	3.299 76 E-24	1.987 17	1	2.083 58 E+4	1.488 68 E+4
1 MHz	6.626 2 E-28	3.990 40 E-4	3.335 64 E-5	4.135 72 E-9	1.583 70 E-28	9.537 34 E-5	4.799 43 E-5	1	7.144 84 E-1
1 G	9.274 1 E-28	5.585 03 E-4	4.668 60 E-5	5.788 38 E-9	2.216 56 E-28	1.334 85 E-4	6.717 36 E-5	1.399 61	1

1.3 Atomic weights and their reciprocals

The following atomic weights apply to elements of terrestrial origin and are based on the scale $^{12}C = 12$. The table incorporates the revised values[3] recommended by the 1969 Atomic Weights Commission of the International Union of Pure and Applied Chemistry. Values are reliable to ±1 in the last digit quoted, except where this is given in small type for which the reliability is ±3. References[3,4] should be consulted for the discussion of the sources of error and the variability of isotopic composition.

An asterisk denotes a radioactive element; if, for such an element, there is no entry under 'naturally occurring isotopes' the atomic weight refers to the most commonly available long-lived isotope. If the figure is in parentheses, it is the mass or mass number only of a common isotope, albeit the element is not yet available in this isotopically pure form.

The naturally occurring isotopes are denoted by the mass number (in bold face type) followed by the natural abundance expressed as a percentage and the nuclear spin quantum number, I.

Element	Atomic symbol	Atomic number	Atomic weight	1000 / Atomic weight	Naturally occurring isotopes (mass no., abundance, nuclear spin)
Actinium*	Ac	89	(227.027 8)	(4.404 747)	
Aluminium	Al	13	26.981 5	37.062 4	**27** 100 $\frac{5}{2}$
Americium*	Am	95	(243.061 4)	(4.114 187)	
Antimony	Sb	51	121.75	8.213 5	**121** 57.25 $\frac{5}{2}$; **123** 42.75 $\frac{7}{2}$
Argon	Ar	18	39.948	25.032	**36** 0.337 0; **38** 0.063 0; **40** 99.60 0
Arsenic	As	33	74.921 6	13.347 28	**75** 100 $\frac{3}{2}$
Astatine*	At	85	(210)		
Barium	Ba	56	137.34	7.281 2	**130** 0.101 0; **132** 0.097 0; **134** 2.42 0; **135** 6.59 $\frac{3}{2}$; **136** 7.81 0; **137** 11.32 $\frac{3}{2}$; **138** 71.66 0
Berkelium*	Bk	97	(247.070 2)	(4.047 433)	
Beryllium	Be	4	9.012 18	110.960 9	**9** 100 $\frac{3}{2}$
Bismuth	Bi	83	208.980 6	4.785 133	**209** 100 $\frac{9}{2}$
Boron	B	5	10.81	92.51	**10** 19.6 3; **11** 80.4 $\frac{3}{2}$
Bromine	Br	35	79.904	12.515 0	**79** 50.54 $\frac{3}{2}$; **81** 49.46 $\frac{3}{2}$
Cadmium	Cd	48	112.40	8.896 8	**106** 1.22 0; **108** 0.88 0; **110** 12.39 0; **111** 12.75 $\frac{1}{2}$; **112** 24.07 0; **113** 12.26 $\frac{1}{2}$; **114** 28.86 0; **116** 7.58 0
Caesium	Cs	55	132.905 5	7.524 14	**133** 100 $\frac{7}{2}$
Calcium	Ca	20	40.08	24.950	**40** 96.97 0; **42** 0.64 0; **43** 0.145 $\frac{7}{2}$; **44** 2.06 0; **46** 0.0033 0; **48** 0.18 0
Californium*	Cf	98	(251)		
Carbon	C	6	12.011	83.257	**12** 98.89 0; **13** 1.11 $\frac{1}{2}$
Cerium	Ce	58	140.12	7.136 7	**136** 0.193 0; **138** 0.250 0; **140** 88.48 0; **142** 11.07 0
Chlorine	Cl	17	35.453	28.206	**35** 75.53 $\frac{3}{2}$; **37** 24.47 $\frac{3}{2}$
Chromium	Cr	24	51.996	19.232 2	**50** 4.31 0; **52** 83.76 0; **53** 9.55 $\frac{3}{2}$; **54** 2.38 0
Cobalt	Co	27	58.933 2	16.968 33	**59** 100 $\frac{7}{2}$
Copper	Cu	29	63.546	15.736 6	**63** 69.09 $\frac{3}{2}$; **65** 30.91 $\frac{3}{2}$
Curium*	Cm	96	(247)		
Dysprosium	Dy	66	162.50	6.153 8	**156** 0.052 0; **158** 0.090 0; **160** 2.29 0; **161** 18.88 $\frac{5}{2}$; **162** 25.53 0; **163** 24.97 $\frac{5}{2}$; **164** 28.18 0
Einsteinium*	Es	99	(254.088 1)	(3.935 643)	

Element	Atomic symbol	Atomic number	Atomic weight	1000 / Atomic weight	Naturally occurring isotopes (mass no., abundance, nuclear spin)
Erbium	Er	68	167.26	5.9787	**162** 0.136 0; **164** 1.56 0; **166** 33.41 0; **167** 22.94 $\frac{7}{2}$; **168** 27.07 0; **170** 14.88 0
Europium	Eu	63	151.96	6.5807	**151** 47.82 $\frac{5}{2}$; **153** 52.18 $\frac{5}{2}$
Fermium*	Fm	100	(253)		
Fluorine	F	9	18.9984	52.6360	**19** 100 $\frac{1}{2}$
Francium*	Fr	87	(223.0198)	(4.483968)	
Gadolinium	Gd	64	157.25	6.3593	**152** 0.200 0; **154** 2.15 0; **155** 14.73 $\frac{3}{2}$; **156** 20.47 0; **157** 15.68 $\frac{3}{2}$; **158** 24.87 0; **160** 21.90 0
Gallium	Ga	31	69.72	14.343	**69** 60.4 $\frac{3}{2}$; **71** 39.6 $\frac{3}{2}$
Germanium	Ge	32	72.59	13.776	**70** 20.52 0; **72** 27.43 0; **73** 7.76 $\frac{9}{2}$; **74** 36.54 0; **76** 7.76 0
Gold	Au	79	196.9665	5.077005	**197** 100 $\frac{3}{2}$
Hafnium	Hf	72	178.49	5.6026	**174** 0.18 0; **176** 5.20 0; **177** 18.50 $\frac{7}{2}$; **178** 27.14 0; **179** 13.75 $\frac{9}{2}$; **180** 35.24 0
Helium	He	2	4.00260	249.8376	**3** 0.00013 $\frac{1}{2}$; **4** 99.99987 0
Holmium	Ho	67	164.9303	6.063167	**165** 100 $\frac{7}{2}$
Hydrogen	H	1	1.0080	992.06	**1** 99.985 $\frac{1}{2}$; **2** 0.015 1
Indium	In	49	114.82	8.7093	**113** 4.28 $\frac{9}{2}$; **115** 95.72 $\frac{9}{2}$
Iodine	I	53	126.9045	7.879941	**127** 100 $\frac{5}{2}$
Iridium	Ir	77	192.22	5.2024	**191** 37.3 $\frac{3}{2}$; **193** 62.7 $\frac{3}{2}$
Iron	Fe	26	55.847	17.9061	**54** 5.82 0; **56** 91.66 0; **57** 2.19 $\frac{1}{2}$; **58** 0.33 0
Krypton	Kr	36	83.80	11.933	**78** 0.35 0; **80** 2.27 0; **82** 11.56 0; **83** 11.55 $\frac{9}{2}$; **84** 56.90 0; **86** 17.37 0
Lanthanum	La	57	138.9055	7.199139	**138** 0.089 5; **139** 99.911 $\frac{7}{2}$
Lawrencium*	Lw	103	(257)		
Lead	Pb	82	207.2	4.826	**204** 1.48 0; **206** 23.6 0; **207** 22.6 $\frac{1}{2}$; **208** 52.3 0
Lithium	Li	3	6.941	144.07	**6** 7.42 1; **7** 92.58 $\frac{3}{2}$
Lutetium	Lu	71	174.97	5.7153	**175** 97.41 $\frac{7}{2}$; **176** 2.59 7
Magnesium	Mg	12	24.305	41.144	**24** 78.70 0; **25** 10.13 $\frac{5}{2}$; **26** 11.17 0
Manganese	Mn	25	54.9380	18.20234	**55** 100 $\frac{5}{2}$
Mendelevium*	Md	101	(256)		
Mercury	Hg	80	200.59	4.9853	**196** 0.146 0; **198** 10.02 0; **199** 16.84 $\frac{1}{2}$; **200** 23.13 0; **201** 13.22 $\frac{3}{2}$; **202** 29.80 0; **204** 6.85 0
Molybdenum	Mo	42	95.94	10.423	**92** 15.84 0; **94** 9.04 0; **95** 15.72 $\frac{5}{2}$; **96** 16.53 0; **97** 9.46 $\frac{5}{2}$; **98** 23.78 0; **100** 9.13 0
Neodymium	Nd	60	144.24	6.9329	**142** 27.11 0; **143** 12.17 $\frac{7}{2}$; **144** 23.85 0; **145** 8.30 $\frac{7}{2}$; **146** 17.22 0; **148** 5.73 0; **150** 5.62 0
Neon	Ne	10	20.179	49.556	**20** 90.92 0; **21** 0.257 $\frac{3}{2}$; **22** 8.82 0
Neptunium*	Np	93	237.0482	4.218551	

Element	Atomic symbol	Atomic number	Atomic weight	$\frac{1000}{\text{Atomic weight}}$	Naturally occurring isotopes (mass no., abundance, nuclear spin)
Nickel	Ni	28	58.71	17.033	**58** 67.88 0; **60** 26.23 0; **61** 1.19 $\frac{3}{2}$; **62** 3.66 0; **64** 1.08 0
Niobium	Nb	41	92.9064	10.76352	**93** 100 $\frac{9}{2}$
Nitrogen	N	7	14.0067	71.39440	**14** 99.63 1; **15** 0.37 $\frac{1}{2}$
Nobelium*	No	102	(253)		
Osmium	Os	76	190.2	5.258	**184** 0.018 0; **186** 1.59 0; **187** 1.64 $\frac{1}{2}$; **188** 13.3 0; **189** 16.1 $\frac{3}{2}$; **190** 26.4 0; **192** 41.0 0
Oxygen	O	8	15.9994	62.5023	**16** 99.759 0; **17** 0.037 $\frac{5}{2}$; **18** 0.204 0
Palladium	Pd	46	106.4	9.398	**102** 0.96 0; **104** 10.97 0; **105** 22.23 $\frac{5}{2}$ **106** 27.33 0; **108** 26.71 0; **110** 11.81 0
Phosphorus	P	15	30.9738	32.2853	**31** 100 $\frac{1}{2}$
Platinum	Pt	78	195.09	5.1258	**190** 0.0127 0; **192** 0.78 0; **194** 32.9 0; **195** 33.8 $\frac{1}{2}$; **196** 25.3 0; **198** 7.21 0
Plutonium*	Pu	94	(242.0587)	(4.131229)	
Polonium*	Po	84	(209.9829)	(4.762293)	
Potassium	K	19	39.102	25.5741	**39** 93.10 $\frac{3}{2}$; **40** 0.0118 4; **41** 6.88 $\frac{3}{2}$
Praseodymium	Pr	59	140.9077	7.096844	**141** 100 $\frac{5}{2}$
Promethium*	Pm	61	(147)		
Protactinium*	Pa	91	231.0359	4.328332	
Radium*	Ra	88	226.0254	4.424281	
Radon*	Rn	86	(222.0175)	(4.504149)	
Rhenium	Re	75	186.2	5.371	**185** 37.07 $\frac{5}{2}$; **187** 62.93 $\frac{5}{2}$
Rhodium	Rh	45	102.9055	9.717653	**103** 100 $\frac{1}{2}$
Rubidium	Rb	37	85.4678	11.70031	**85** 72.15 $\frac{5}{2}$; **87** 27.85 $\frac{3}{2}$
Ruthenium	Ru	44	101.07	9.8941	**96** 5.51 0; **98** 1.87 0; **99** 12.72 $\frac{3}{2}$ **100** 12.62 0; **101** 17.07 $\frac{5}{2}$; **102** 31.61 0; **104** 18.58 0
Samarium	Sm	62	150.4	6.649	**144** 3.09 0; **147** 14.97 $\frac{7}{2}$; **148** 11.24 0; **149** 13.83 $\frac{7}{2}$; **150** 7.44 0; **152** 76.72 0; **154** 22.71 0
Scandium	Sc	21	44.9559	22.2440	**45** 100 $\frac{7}{2}$
Selenium	Se	34	78.96	12.665	**74** 0.87 0; **76** 9.02 0; **77** 7.58 $\frac{1}{2}$; **78** 23.52 0; **80** 49.82 0; **82** 9.19 0
Silicon	Si	14	28.086	35.605	**28** 92.21 0; **29** 4.70 $\frac{1}{2}$; **30** 3.09 0
Silver	Ag	47	107.868	9.27059	**107** 51.82 $\frac{1}{2}$; **109** 48.18 $\frac{1}{2}$
Sodium	Na	11	22.9898	43.4975	**23** 100 $\frac{3}{2}$
Strontium	Sr	38	87.62	11.4129	**84** 0.56 0; **86** 9.86 0; **87** 7.02 $\frac{9}{2}$; **88** 82.56 0
Sulphur	S	16	32.06	31.192	**32** 95.0 0; **33** 0.76 $\frac{3}{2}$; **34** 4.22 0; **36** 0.014 0
Tantalum	Ta	73	180.9479	5.526452	**180** 0.0123 ?; **181** 99.988 $\frac{7}{2}$
Technetium*	Tc	43	98.9062	10.11059	
Tellurium	Te	52	127.60	7.8370	**120** 0.089 0; **122** 2.46 0; **123** 0.87 $\frac{1}{2}$; **124** 4.61 0; **125** 6.99 $\frac{1}{2}$; **126** 18.71 0; **128** 31.79 0; **130** 34.48 0

Element	Atomic symbol	Atomic number	Atomic weight	$\frac{1000}{\text{Atomic weight}}$	Naturally occurring isotopes (mass no., abundance, nuclear spin)
Terbium	Tb	65	158.9254	6.292260	**159** 100 $\frac{3}{2}$
Thallium	Tl	81	204.37	4.8931	**203** 29.50 $\frac{1}{2}$; **205** 70.50 $\frac{1}{2}$
Thorium*	Th	90	232.0381	4.309637	
Thulium	Tm	69	168.9342	5.919464	**169** 100 $\frac{1}{2}$
Tin	Sn	50	118.69	8.4253	**112** 0.96 0; **114** 0.66 0; **115** 0.35 $\frac{1}{2}$; **116** 14.30 0; **117** 7.61 $\frac{1}{2}$; **118** 24.03 0; **119** 8.58 $\frac{1}{2}$; **120** 32.85 0; **122** 4.92 0; **124** 5.94 0
Titanium	Ti	22	47.90	20.877	**46** 7.93 0; **47** 7.28 $\frac{5}{2}$; **48** 73.94 0; **49** 5.51 $\frac{7}{2}$; **50** 5.34 0
Tungsten	W	74	183.85	5.4392	**180** 0.14 0; **182** 26.41 0; **183** 14.40 $\frac{1}{2}$; **184** 30.64 0; **186** 28.41 0
Uranium*	U	92	238.029	4.20117	**234** 0.0057 0; **235** 0.72 $\frac{7}{2}$; **238** 99.27 0
Vanadium	V	23	50.9414	19.63040	**50** 0.24 6; **51** 99.76 $\frac{7}{2}$
Xenon	Xe	54	131.30	7.6161	**124** 0.096 0; **126** 0.090 0; **128** 1.92 0; **129** 26.44 $\frac{1}{2}$; **130** 4.08 0; **131** 21.18 $\frac{3}{2}$ **132** 26.89 0; **134** 10.44 0; **136** 8.87 0
Ytterbium	Yb	70	173.04	5.7790	**168** 0.135 0; **170** 3.03 0; **171** 14.31 $\frac{1}{2}$; **172** 21.82 0; **173** 16.13 $\frac{5}{2}$; **174** 31.84 0; **176** 12.73 0
Yttrium	Y	39	88.9059	11.24785	**89** 100 $\frac{1}{2}$
Zinc	Zn	30	65.37	15.298	**64** 48.89 0; **66** 27.81 0; **67** 4.11 $\frac{5}{2}$; **68** 18.57 0; **70** 0.62 0
Zirconium	Zr	40	91.22	10.963	**90** 51.46 0; **91** 11.23 $\frac{5}{2}$; **92** 17.11 0; **94** 17.40 0; **96** 2.80 0

2 Geometrical data

2.1 Trigonometric formulae

$\sin(A \pm B) = \sin A \cos B \pm \cos A \sin B$

$\cos(A \pm B) = \cos A \cos B \mp \sin A \sin B$

$\sin A \pm \sin B = 2 \sin \frac{1}{2}(A \pm B) \cos \frac{1}{2}(A \mp B)$

$\cos A + \cos B = 2 \cos \frac{1}{2}(A + B) \cos \frac{1}{2}(A - B)$

$\cos A - \cos B = 2 \sin \frac{1}{2}(A + B) \sin \frac{1}{2}(B - A)$

$\sin 2A = 2 \sin A \cos A$

$\cos 2A = \cos^2 A - \sin^2 A = 1 - 2 \sin^2 A = 2 \cos^2 A - 1$

$\sin 3A = 3 \sin A - 4 \sin^3 A$

$\cos 3A = 4 \cos^3 A - 3 \cos A$

$\sin A = \sin[n\pi + (-1)^n A]$ } n is any positive or

$\cos A = \cos[2n\pi \pm A]$ } negative integer

$e^{ix} = \cos x + i \sin x$

$e^{-ix} = \cos x - i \sin x$

$e^{ix} + e^{-ix} = 2 \cos x$

$\cos x = \frac{1}{2}(e^{ix} + e^{-ix})$

$\sin x = \frac{1}{2i}(e^{ix} - e^{-ix})$

$\cosh x = \cos ix = \frac{1}{2}(e^{x} + e^{-x})$

$\sinh x = \frac{1}{i} \sin ix = \frac{1}{2}(e^{x} - e^{-x})$

$(\cos x + i \sin x)^n = \cos nx + i \sin nx$

The tetrahedral angle, ϕ, is the angle subtended at the centroid of a regular tetrahedron by one of its edges.

$\cos \phi = -\frac{1}{3} = -0.333\,333$

$\sin \phi = \frac{2}{3}\sqrt{2} = 0.942\,809$

$\tan \phi = -2\sqrt{2} = -2.828\,427$

$\cos \frac{\phi}{2} = \frac{1}{\sqrt{3}} = 0.577\,350$

$\sin \frac{\phi}{2} = \sqrt{\frac{2}{3}} = 0.816\,497$

$\tan \frac{\phi}{2} = \sqrt{2} = 1.414\,214$

$\phi = 109.471\,220\,634\ldots°$

The 'golden section', τ, arises in five-fold symmetry, and will be used in these tables to express trigonometric functions of the angle $\frac{2\pi}{5}$ (72°) and its multiples. The 'golden section' of classical antiquity by definition satisfies the equation:

$$\tau^2 = \tau + 1 \quad \text{or} \quad \frac{1}{\tau} = \tau - 1$$

of which the positive root is:

$$\tau = \tfrac{1}{2}(\sqrt{5} + 1) = 1.618\,033\,988\,7\ldots .$$

In a regular pentagon it may be readily shown that:

$$(1 + 2 \cos \tfrac{2\pi}{5})^2 = 2(1 + \cos \tfrac{2\pi}{5})$$

whence $2 \cos \frac{2\pi}{5} = \frac{1}{2}(\sqrt{5} - 1)$.

Thus: $2 \cos \frac{2\pi}{5} = -2 \cos \frac{3\pi}{5} = \tau - 1$ and $2 \cos \frac{\pi}{5} = -2 \cos \frac{4\pi}{5} = \tau$.

The relation $\tau^2 = \tau + 1$ enables the characters of five-fold symmetry operations to be manipulated without numerical evaluation.

2.2 Convex polyhedra

The following table summarises some properties of convex polyhedra selected for their importance in chemistry. The solids are listed in the order of increasing numbers of vertices, and are denoted by the symbols of Cundy and Rollett[5] (whose book should be consulted for more details of these and other uniform polyhedra): for the facially regular isogonal solids the types of regular face meeting at a vertex are written as the numbers of sides in each such face, and the number of faces of each type is written as an index. For the vertically regular isohedral solids the symbol is prefixed by V, the types of regular vertex around a face are written as the orders of the vertices, and the number of vertices of each type is written as an index. The suffices to the numbers of faces (fifth column) are the numbers of sides in each face.

A and V denote the area and volume, respectively, of a solid having unit edge. All the solids listed obey Euler's theorem which may be expressed by the equation:

Faces + Vertices = Edges + 2 .

Name	Symbol	Point group	Number of vertices	Number of faces	Number of edges	Angle subtended by edge at centre (θ)	Edge / Circumrad. ($2 \sin \frac{1}{2}\theta$)	Dihedral angles	Comments
Tetrahedron	3^3	T_d	4	4_3	6	109.47	1.6329	70.53	Regular, $A = \sqrt{3}$, $V = \frac{\sqrt{2}}{12}$
Octahedron	3^4	O_h	6	8_3	12	90.00	1.4142	109.47	Regular, $A = 2\sqrt{3}$, $V = \frac{\sqrt{2}}{3}$
Cube	4^3	O_h	8	6_4	12	70.53	1.1547	90.00	Regular, $A = 6$, $V = 1$
Trigonal dodecahedron	-	D_{2d}	8	12_3	18	-	-	-	
Symmetrically tricapped trigonal prism	-	D_{3h}	9	14_3	21	-	-	-	
Symmetrically bicapped square antiprism	-	D_{4d}	10	16_3	24	-	-	-	
Icosahedron	3^5	I_h	12	20_3	30	63.43	1.0515	138.19	Regular, $A = 5\sqrt{3}$, $V = \frac{10}{3}\cos^2 36$
Truncated tetrahedron	3.6^2	T_d	12	$4_3 + 4_6$	18	50.47	0.8519	70.53 109.47	Isogonal
Cuboctahedron	$(3.4)^2$	O_h	12	$8_3 + 6_4$	24	60.00	1.0000	125.27	Isogonal
Rhombic dodecahedron	$V(3.4)^2$	O_h	14	12_4	24	55.87	0.9369	120.00	Isohedral, space filling solid in ccp. $A = 4\sqrt{2}$ $V = \frac{16}{3}\sqrt{3}$
Dodecahedron	5^3	I_h	20	12_5	30	41.80	0.7136	116.57	Regular, $A = 15 \cot 36$ $V = 5 \cot^2 36 \cos 36$
Truncated octahedron	4.6^2	O_h	24	$6_4 + 8_6$	36	36.87	0.6325	109.47 125.27	Isogonal

Name	Symbol	Point group	Number of vertices	Number of faces	Number of edges	Angle subtended by edge at centre (θ)	$\frac{\text{Edge}}{\text{Circumrad.}}$ ($2 \sin \frac{1}{2}\theta$)	Dihedral angles	Comments
Truncated cube	3.8^2	$\mathbf{O_h}$	24	8_3+6_8	36	32.65	0.5622	$\begin{cases} 90.00 \\ 125.27 \end{cases}$	Isogonal
n-gonal archimedian prism	$4^2.n$	$\mathbf{D_{nh}}$	2n	2_n+n_4	3n	$\tan^{-1}(\sin\frac{\pi}{n})$	$\frac{2\sin\frac{\pi}{n}}{\sqrt{(1+\sin^2\frac{\pi}{n})}}$	$\begin{cases} 90.00 \\ \pi(1-\frac{2}{n}) \end{cases}$	Isogonal
n-gonal archimedian antiprism	$3^3.n$	$\mathbf{D_{nd}}$	2n	2_n+2n_3	4n	$\tan^{-1}(2\sin\frac{\pi}{2n})$	$\frac{4\sin\frac{\pi}{2n}}{\sqrt{(3-2\cos\frac{\pi}{n})}}$	$\begin{cases} \sec^{-1}(-\sqrt{3}[\operatorname{cosec}\frac{\pi}{n}+\cot\frac{\pi}{n}]) \\ \cos^{-1}(\frac{1}{3}[1-4\cos\frac{\pi}{n}]) \end{cases}$	
n-gonal bipyramid	$V.4^2.n$	$\mathbf{D_{nh}}$	n+2	$2n_3$	3n	-	-	-	Isohedral

3 Principles of symmetry

3.1 Symmetry elements and symmetry operations

Operation: A transformation of coordinates; alternatively, a transformation of the molecule to a new position.

Symmetry operation: An operation that interchanges indistinguishable particles; alternatively, an operation (not necessarily a physically feasible one) that carries a molecule into a position indistinguishable from (or equivalent to) the original position. It sends a molecule into self-coincidence.

Symmetry element: A geometrical entity (point, line or plane) within a molecule, with respect to which certain symmetry operations may be carried out.

Proper operations: Such symmetry operations may be viewed as pure rotations about a specified axis: they are physically feasible and do not change the chirality ('handedness') of a molecule.

Improper operations: These may be regarded as rotation-reflection operations: they are not physically feasible and they do change the chirality of the molecule.

Sense of rotation: The conventional definition of a positive rotation is the sense of rotation of a corkscrew advancing along the positive direction of the specified axis.

Symmetry element		Corresponding operations	Definition of symmetry operation	Order of Element	Relationship between operations
Symbol	Name				
-	-	E	The identity	1	$E^k = E$
		PROPER OPERATIONS			
C_n	n-fold rotation axis. (Assumed to lie in the z direction, except in cubic and icosahedral point groups.)	$C_n^1, C_n^2, \ldots C_n^k, \ldots C_n^n$	Rotation through an angle $2\pi k/n$ about the axis (when k=1, the superscript is usually omitted, i.e. $C_n^1 = C_n$)	n	$C_n^n = E$ $C_n^{n+k} = C_n^k$ $(C_n^k)^{-1} = C_n^{-k} = C_n^{n-k}$ Factoring: $C_n^k = C_{n/k}^1$ if n/k is integral (operations are always writt en in their simplest form)
Special cases:					
C'_2, C''_2	2-fold rotation axis perpendicular to the principal C_n axis	C'_2, C''_2	Rotation through π about the axis.	2	
$C_n(x)$ (etc)	n-fold rotation axis in the x direction (etc)				
C_∞	Infinite-fold rotation axis in the z direction (present in all linear molecules)	C_∞^ϕ	Rotation through an arbitrary angle ϕ about the axis.	∞	$(C_\infty^\phi)^{-1} = C_\infty^{-\phi}$
		IMPROPER OPERATIONS			
S_n	n-fold rotation-reflection axis. (Or 'alternating axis'. Assumed to lie in the z direction except in the cubic and icosahedral point groups)	$S_n^1, S_n^2, \ldots S_n^k, \ldots S_n^n$ $S_n^1, S_n^2, \ldots S_n^k, \ldots S_n^n$ $\ldots S_n^{n+k}, \ldots S_n^{2n}$	Rotation about the axis through $2\pi k/n$, combined with reflection k times in a plane normal to the axis (i.e. if k is odd, net result is one reflection; if k even, no reflection and the operation becomes proper)	n (n even) 2n (n odd)	$S_n^n = E$ (n even) $S_n^{2n} = E$ (n odd) $S_n^k = C_n^k$ (k even) $S_n^{n+k} = C_n^k$ (n and k odd)
Special cases:					
i ($=S_2$)	inversion centre	i ($= S_2^1$)	Inversion of all points through the origin of coordinates.	2	$i^2 = E$ $i = S_{2n}^n$ (n odd)

Symmetry element		Corresponding operations	Definition of symmetry operation	Order of Element	Relationship between operations
Symbol	Name				
$\sigma(=S_1)$	mirror plane	$\sigma(=S_1^1)$	Reflection of all points in a plane.	2	$\sigma^2 = E$
σ_v	a mirror plane containing the principal axis (v = 'vertical')	σ_v	Reflection in the plane specified.		
σ_h	a mirror plane normal to the principal axis (h = 'horizontal')	σ_h	Reflection in the plane specified.		$\sigma_h = S_n^n$ (n odd)
σ_d	a dihedral mirror plane containing the principal axis and bisecting the angles between C_2' axes	σ_d	Reflection in the plane specified.		

3.2 Definitions in group theory

Group criteria

A group is a collection of operations which, when a rule of combination of these is defined, satisfy four conditions.

(i) If any two operations, P and Q, are members of the group, then the combination of these $P.Q = R$ must also be a member of the group. This includes the case where $P = Q$.

(ii) Under the rule of combination the associative law must hold

i.e. $(P.Q).R = P.(Q.R)$.

(iii) One of the operations of the group must be the identity operation E which commutes with all the other operations and leaves them unchanged. Thus,

$E.R = R.E = R$.

(iv) Each member of the group must possess an inverse which is also a member of the group. Thus,

$R^{-1}.R = E = R.R^{-1}$.

Subgroups

A subgroup is a set of elements within a group which, on their own constitute a group.

Conjugate operations

If P, Q and X are operations of a particular group, then P and Q are said to be conjugate if a similarity transform of P by X yields Q, thus:

$X^{-1}.P.X = Q$.

Order of a group

The number of elements in a group is known as the order of the group, (symbol h).

Classes of operations

All the operations conjugate to P, $X^{-1}.P.X$ (generated by running X through the whole group), constitute a class of the group.

Any two operations of a given class are conjugate to one another with respect to some operation of the group.

Corresponding operations from different symmetry elements are in the same class if the elements themselves are symmetrically equivalent, i.e. they may be interchanged by another operation of the group.

Abelian groups

A group in which all operators commute is said to be **abelian**. Every operator in such a group is in a class by itself.

Symmetrically equivalent atoms

Two atoms in a molecule are said to be symmetrically equivalent if one may be replaced by the other in the performance of some symmetry operation.

The nuclei of symmetrically equivalent atoms are **isochronous**, i.e. they have the same magnetic resonance frequency.

Chemically equivalent atoms

Two atoms in a molecule are chemically equivalent if one may be replaced by the other in the performance of a **proper** symmetry operation.

Two atoms which are symmetrically equivalent by virtue of an improper symmetry operation are chemically equivalent with respect to optically inactive (achiral) reagents only.

Magnetically equivalent nuclei

Two nuclei in a molecule are magnetically equivalent if they are symmetrically equivalent by virtue of a symmetry operation that does **not** permute any of the nuclei to which they may be spin coupled.

Generators of a group

A set, P, of elements of a finite group, G, is a system of generators of G, if every element of G can be expressed as a product containing only integer powers (positive or negative) of elements of P.

The elements of P are conveniently chosen to be independent (i. e. none can be expressed in terms of the others); the group multiplication table is implicit in the defining relations for the group generators.

Normal subgroups

A normal (or invariant) subgroup, **H**, of a group, **G**, is self-conjugate, i. e. it commutes with every element, **X**, of **G**:

$$\mathbf{X}^{-1}.\mathbf{H}.\mathbf{X} = \mathbf{H}$$

H is composed of complete classes of **G**.

If there exists no other normal subgroup, **A**, of **G**, such that **H** is a normal subgroup of **A**, then **H** is said to be a **maximal** normal subgroup of **G**.

3.3 The point groups and their properties

Notation

Each symmetry point group is represented by a symbol consisting of a capital letter and, very often, one or two suffixes. The capitals, **C S D T O I** are used to describe the pure rotations of the group; a subscript integer **n**, shows the order of the principal axis, (if any). It is sufficient to define only one improper operation of a group - the rest are obtained by multiplying all the rotations in turn by the defined improper operation. The following suffices are added to indicate the type of operation present:

h a σ_h mirror plane;

v a σ_v mirror plane, but not σ_h;

d a σ_d mirror plane.

A few of the point groups have acquired commonly used alternative symbols; these are included in the following table, which is a complete list of the types of symmetry point group.

Successive, and exhaustive application of the generating operations within the limits of the defining relations yields all the symmetry operations of the group, and the group multiplication table. Note: even the most complicated groups require only three generators. In the axial groups, the principal axis C_n is taken to lie in the z direction. **R** denotes a pure rotation group.

Type		Symbol	Generating operations	Symmetry elements	Order	Number of classes and irreducible representations	Comments
Non-axial		$\mathbf{C_1}$	E	None	1	1	No symmetry
		$\mathbf{C_s}$	σ	σ_h	2	2	$\mathbf{C_s = C_{1h} = C_{1v} = S_1}$
		$\mathbf{C_i}$	i	i	2	2	$\mathbf{C_i = S_2}$
Axial: Cyclic	R	$\mathbf{C_n}$	C_n	C_n	n	n	n = 2, 3, 4, ...
		$\mathbf{S_{2n}}$	S_{2n}	C_n, S_{2n}	2n	2n	$\mathbf{S_6 = C_{3i}}$
		$\mathbf{C_{nh}}$	C_n, σ_h	C_n, σ_h, S_n	2n	2n	
		$\mathbf{C_{nv}}$	C_n, σ_v	C_n, $n\sigma_v$	2n	$\frac{1}{2}(n+3)$	if n odd
Dihedral	R	$\mathbf{D_n}$	C_n, C'_2	C_n, nC'_2	2n	$\frac{1}{2}(n+6)$	if n even; $\mathbf{D_2 = V}$
		$\mathbf{D_{nh}}$	C_n, C'_2, σ_h	C_n, nC'_2, S_n, σ_h, $n\sigma_v$	4n	n+3 n+6	if n odd if n even; $\mathbf{D_{2h} = V_h}$
		$\mathbf{D_{nd}}$	C_n, C'_2, σ_d	C_n, nC'_2, $n\sigma_d$, S_{2n}	4n	n+3	$\mathbf{D_{nd} = S_{2nv}}$; $\mathbf{D_{2d} = V_d}$
Linear		$\mathbf{C_{\infty v}}$	C^{ϕ}_{∞}, σ_v	C_{∞}, $\infty\sigma_v$	∞	∞	
		$D_{\infty h}$	C^{ϕ}_{∞}, C'_2, σ_h	C_{∞}, $\infty\sigma_v$, S_{∞}, $\infty C'_2$	∞	∞	
Cubic	R	$\mathbf{T}$	$C^{\dagger}_3$, $C_2(z)$	$4C_3$, $3C_2$	12	4	Rotations of the regular tetrahedron
		$\mathbf{T_h}$	$C^{\dagger}_3$, $C_2(z)$, i	$4C_3$, $3C_2$, $4S_6$, $3\sigma_v$	24	8	$\mathbf{T_h = C_i \times T}$
		$\mathbf{T_d}$	$C^{\dagger}_3$, $S^3_4(z)$	$4C_3$, $3C_2$, $3S_4$, $6\sigma_d$	24	5	Full symmetry of the regular tetrahedron
	R	$\mathbf{O}$	$C^{\dagger}_3$, $C_4(z)$	$4C_3$, $3C_4$, $6C_2$	24	5	Rotations of the regular octahedron
		$\mathbf{O_h}$	$C^{\dagger}_3$, $C_4(z)$, i	$4C_3$, $3C_4$, $6C_2$, $3S_4$, $4S_6$, $3\sigma_h$, $6\sigma_d$	48	10	Full symmetry of the regular octahedron
Icosahedral	R	$\mathbf{I}$	$C^{\ddagger}_3$, $C_5(z)$	$6C_5$, $10C_3$, $15C_2$	60	5	Rotations of the regular icosahedron
		$\mathbf{I_h}$	$C^{\ddagger}_3$, $C_5(z)$, i	$6C_5$, $10C_3$, $15C_2$, $12S_{10}$, $10S_6$, 15σ	120	10	Full symmetry of the regular icosahedron

† C_3 axis inclined at an angle of 54.74° to the $C_2(z)$ axis; ‡ C_3 axis inclined at an angle of 37.38° to the $C_5(z)$ axis.

Defining relations for the point groups in terms of their generators

Symbol	Defining relations of generators
$\mathbf{C_1}$	$E^2 = E$
$\mathbf{C_s}$	$\sigma^2 = E$
$\mathbf{C_i}$	$i^2 = E$
$\mathbf{C_n}$	$C_n^n = E$
$\mathbf{S_{2n}}$	$S_{2n}^{2n} = E$
$\mathbf{C_{nh}}$	$C_n^n = \sigma_h^2 = E;\ C_n.\sigma_h = \sigma_h.C_n$ [if n is odd $S_n^{2n} = E$]
$\mathbf{C_{nv}}$	$C_n^n = \sigma_v^2 = (\sigma_v.C_n)^2 = E$
$\mathbf{D_n}$	$C_n^n = (C'_2)^2 = (C'_2.C_n)^2 = E$
$\mathbf{D_{nh}}$	$C_n^n = (C'_2)^2 = \sigma_h^2 = (C'_2.C_n)^2 = E;\ C_n.\sigma_h = \sigma_h.C_n;\ C'_2.\sigma_h = \sigma_h.C'_2$ [if n is odd $S_n^{2n} = (C'_2)^2 = (C'_2.S_n)^2 = E$]
$\mathbf{D_{nd}}$	$C_n^n = (C'_2)^2 = \sigma_d^2 = (C'_2.C_n)^2 = (\sigma_d.C_n)^2 = E;\ (\sigma_d.C'_2)^2 = C_n$ [or, $S_{2n}^{2n} = (C'_2)^2 = (C'_2.S_{2n})^2 = E$]
$\mathbf{T}$	$C_3^3 = C_2^2 = (C_2.C_3)^3 = E$
$\mathbf{T_h}$	$C_3^3 = C_2^2 = (C_2.C_3)^3 = i^2 = E;\ C_3.i = i.C_3;\ C_2.i = i.C_2$
$\mathbf{T_d}$	$C_3^3 = S_4^4 = (S_4^3.C_3)^2 = E$
$\mathbf{O}$	$C_3^3 = C_4^4 = (C_4.C_3)^2 = E$
$\mathbf{O_h}$	$C_3^3 = C_4^4 = (C_4.C_3)^2 = i^2 = E;\ C_3.i = i.C_3;\ C_4.i = i.C_4$
$\mathbf{I}$	$C_3^3 = C_5^5 = (C_3.C_5)^2 = E$
$\mathbf{I_h}$	$C_3^3 = C_5^5 = (C_3.C_5)^2 = i^2 = E;\ C_3.i = i.C_3;\ C_5.i = i.C_5$

3.4 Isomorphic and direct product groups

Isomorphism: Two groups are isomorphous (~) when there exists a one-to-one correspondence between their operations; they have the same defining relations and the same multiplication and character tables.

Generally: $\mathbf{C_{nv}} \sim \mathbf{D_n};\ \mathbf{D_{nd}} \sim \mathbf{D_{2n}}$.

If **n** is even: $\mathbf{C_n} \sim \mathbf{S_n}$.

If **n** is odd: $\mathbf{D_{nd}} \sim \mathbf{D_{nh}}$.

Special cases: $\mathbf{C_i} \sim \mathbf{C_2} \sim \mathbf{C_s};\ \mathbf{C_6} \sim \mathbf{C_{3h}};\ \mathbf{D_2} \sim \mathbf{C_{2h}};\ \mathbf{O} \sim \mathbf{T_d}$.

Direct product groups: If there are two groups $\mathbf{G_a}$ and $\mathbf{G_b}$, having only the identity in common and possessing elements a_i, $i = 1, \ldots h_a$, and b_j, $j = 1, \ldots h_b$, such that $a_ib_j = b_ja_i$ for all i and j, then the direct product group $\mathbf{G} = \mathbf{G_a} \times \mathbf{G_b}$ is defined as the set of all distinct elements a_ib_j produced for all i and j. $\mathbf{G_a}$ and $\mathbf{G_b}$ are normal subgroups of $\mathbf{G}$. Important special cases arise when one of the product groups is a pure rotation group and the other is $\mathbf{C_s} = (E, \sigma_h)$ or $\mathbf{C_i} = (E, i)$; the character tables are then related by the following direct products.

If n is even:	$C_{nh} = C_i \times C_n$
	$D_{nh} = C_i \times D_n$
If n is odd:	$S_{2n} = C_i \times C_n$
	$C_{nh} = C_s \times C_n$
	$D_{nh} = C_s \times D_n$
	$D_{nd} = C_i \times D_n$
Special cases:	$T_h = C_i \times T$
	$O_h = C_i \times O$
	$I_h = C_i \times I$
	$O(3) = C_i \times R(3)$

In the character tables (§6) of these direct product groups, we have listed the classes of the pure rotation group first, followed by the classes of improper operations in an order determined by the following relationships:

Pure rotation R	$i \times R$	$\sigma_h \times R$
C_1^1	S_2^1 (= i)	S_1^1 (= σ_h)
C_2^1	S_1^1 (= σ_h)	S_2^1 (= i)
C_3^1	S_6^5	S_3^1
C_3^2	S_6^1	S_3^5
C_4^1	S_4^3	S_4^1
C_4^3	S_4^1	S_4^3
C_5^1	S_{10}^7	S_5^1
C_5^2	S_{10}^9	S_5^7
C_5^3	S_{10}^1	S_5^3
C_5^4	S_{10}^3	S_5^9
C_6^1	S_3^5	S_6^1
C_6^5	S_3^1	S_6^5
C_8^1	S_8^5	S_8^1
C_8^3	S_8^7	S_8^3
C_8^5	S_8^1	S_8^5
C_8^7	S_8^3	S_8^7
C_{10}^1	S_5^3	S_{10}^1
C_{10}^3	S_5^9	S_{10}^3
C_{10}^7	S_5^1	S_{10}^7
C_{10}^9	S_5^7	S_{10}^9
C_{12}^1	S_{12}^7	S_{12}^1
C_{12}^5	S_{12}^{11}	S_{12}^5
C_{12}^7	S_{12}^1	S_{12}^7
C_{12}^{11}	S_{12}^5	S_{12}^{11}

The characters, $\chi(C)$, of an irreducible representation of a pure rotation group, $\mathbf{G_R}$, are related to the characters, $\chi_g(C)$, $\chi_g(S)$; $\chi_u(C)$, $\chi_u(S)$, of the corresponding g and u representations of the direct product group $\mathbf{C_i} \times \mathbf{G_R}$ when the proper and improper operations, C and S, are in correspondence (i. e. $S = i \times C$):

$$\chi_g(C) = \chi_g(S) = \chi_u(C) = \chi(C) \quad \text{and} \quad \chi_u(S) = -\chi(C);$$

similarly for the direct product group $\mathbf{C_s} \times \mathbf{G_R}$:

$$\chi'(C) = \chi'(S) = \chi''(C) = \chi(C) \quad \text{and} \quad \chi''(S) = -\chi(C)$$

where $S = \sigma_h \times C$.

3.5 Transformations of coordinates

A general point in 3-space, Cartesian coordinates (x, y, z), may be transformed into another point (x', y', z') by either (a) a rotation (or rotation/reflection) of the coordinate axes, or (b) a rotation (or rotation/reflection) of the point itself. Transformation (b) is the inverse of (a) if the same sense of rotation is preserved.

The matrix R which transforms the vector (x, y, z) into (x', y', z') is orthogonal, in which case its inverse is equal to its transpose:

$$R^{-1} = \tilde{R}\ .$$

Listed below are the matrices corresponding to case (a), symmetry operations applied to the coordinate system. The matrices for case (b) are readily derived by transposition.

Rotation (C_n^k) and rotation-reflection (S_n^k) about the axes x, y and z

$$R(z) = \begin{bmatrix} \cos\alpha & \sin\alpha & 0 \\ -\sin\alpha & \cos\alpha & 0 \\ 0 & 0 & \pm 1 \end{bmatrix}$$

$$R(y) = \begin{bmatrix} \cos\alpha & 0 & -\sin\alpha \\ 0 & \pm 1 & 0 \\ \sin\alpha & 0 & \cos\alpha \end{bmatrix}$$

$$R(x) = \begin{bmatrix} \pm 1 & 0 & 0 \\ 0 & \cos\alpha & \sin\alpha \\ 0 & -\sin\alpha & \cos\alpha \end{bmatrix}$$

$\alpha = 2\pi k/n$. In ± 1, the + sign is taken for proper, $R = C_n^k$ operations, the - sign is taken for improper, $R = S_n^k$ operations.

Dihedral planes and axes

$$\sigma_d = \begin{bmatrix} \cos\beta & \sin\beta & 0 \\ \sin\beta & -\cos\beta & 0 \\ 0 & 0 & 1 \end{bmatrix}$$

($\beta = 2\pi/n$. σ_d makes an angle π/n with the x axis, and contains z; n refers to the order of the principal axis of the $\mathbf{D_{nd}}$ group.)

$$C''_2 = \begin{bmatrix} \cos\beta & \sin\beta & 0 \\ \sin\beta & -\cos\beta & 0 \\ 0 & 0 & -1 \end{bmatrix}$$

($\beta = 2\pi/n$. C''_2 makes an angle π/n with the x axis. n refers to the order of the principal axis of the $\mathbf{D}_{nh}$ group.)

Special axes in the cubic and icosahedral groups

$$C_3^{(xyz)} = \begin{bmatrix} 0 & 1 & 0 \\ 0 & 0 & 1 \\ 1 & 0 & 0 \end{bmatrix}$$

$C_3^{(xyz)}$ makes an angle of 54.74° with the x, y and z axes.

$$C_3^{icos} = \begin{bmatrix} -\frac{1}{2} & \frac{3}{2}\cos\delta & \frac{3}{2}\sin\delta \\ -\frac{3}{2}\cos\delta & 1-\frac{3}{2}\cos^2\delta & \frac{3}{2}\sin\delta\cos\delta \\ \frac{3}{2}\sin\delta & \frac{3}{2}\sin\delta\cos\delta & 1-\frac{3}{2}\sin^2\delta \end{bmatrix}$$

C_3^{icos} lies in the yz plane, and makes an angle δ with the z axis. $\cos\delta = \frac{1}{3}\cot 36° = \frac{1}{3}\cot\frac{\pi}{5}$. $\delta = 37.38°$.

3.6 Properties of irreducible representations

Notation

h	Order of a group, **H**.
d_i	Dimension of the ith irreducible representation, Γ_i of **H**.
$\chi_i(R)$	Character of Γ_i under a symmetry operation of the Rth class of **H**.
$\chi_i(A)$	Character of Γ_i under the Ath symmetry operation of **H**.
g_R	Number of symmetry operations in the Rth class of **H**.
$D_i(A)_{kl}$	The k, l matrix element of Γ_i under the Ath symmetry operation of **H**.
δ_{ij}	Kronecker delta $\begin{Bmatrix} \delta_{ij} = 0 \text{ if } i \neq j \\ \delta_{ij} = 1 \text{ if } i = j \end{Bmatrix}$
$\sum_A$	Summation over all symmetry operations in the group.
$\sum_R$	Summation over all classes in the group.

(1) The number of irreducible representations of a group is equal to the number of classes in the group.

(2) The character of an irreducible representation is the trace of the matrix $D_i(A)$

$$\chi_i(A) = \sum_{k=1}^{k=d_i} D_i(A)_{kk}.$$

(3) In a given representation the characters of all matrices belonging to operations in the same class are equal.

(4) The characters of irreducible representations obey the orthogonality relationship:

$$\sum_R g_R.\ \chi_i(R).\ \chi_j(R)^* = h\delta_{ij}.$$

Properties of characters (real χ)

(5) The sum of the squares of the characters under the identity (E) of the irreducible representations of the group equals the order of the group, (h):

$$\sum_i (\chi_i(E))^2 = h .$$

(6) The sum of the squares of the characters over all classes in any irreducible representation equals the order of the group:

$$\sum_R g_R \cdot \chi_i(R) . \chi_i(R) = h .$$

(7) The characters of two different irreducible representations are orthogonal:

$$\sum_R g_R \cdot \chi_i(R) . \chi_j(R) = 0 \text{ when } i \neq j .$$

The orthogonality of the irreducible matrix representations is expressed by The Great Orthogonality Theorem:

$$\sum_A D_i(A)_{kl} \cdot D_j(A)^*_{mn} = \frac{h}{d_i} \cdot \delta_{ij} \cdot \delta_{km} \cdot \delta_{ln} .$$

Reduction of a representation:

If $\Gamma = \sum_i a_i \Gamma_i$

$$a_i = \frac{1}{h} \sum_R g_R \cdot \chi_i(R)^* . \chi(R) .$$

The projection operator using characters of irreducible representations:

$$P_i = \frac{d_i}{h} \sum_A \chi_i(A)^* . A .$$

The projection operator using irreducible matrix representations:

$$P_i(kl) = \frac{d_i}{h} \sum_A D_i(A)^*_{kl} . A .$$

Use of projection operators

If a set of N symmetrically equivalent elements $\{\phi_n\}$ forms a basis for an N-dimensional reducible representation Γ of **H** (i. e. the $\{\phi_n\}$ may be nuclear displacements or wave functions of atomic orbitals), then the projection operators will provide the orthonormal bases which transform as those irreducible representations, Γ_i, of **H** which are spanned by Γ (i. e. symmetry coordinates or linear combinations of atomic orbitals).

The projection operator using characters, P_i, when applied to an arbitrary element, ϕ_n, of the basis set generates a linear combination:

$$P_i \cdot \phi_n = \sum_{k=1}^{N} C_{ik} \phi_k$$

that constitutes **one** component of a basis for Γ_i. If Γ_i is non-degenerate ($d_i = 1$, $\Gamma_i = A$ or B) this is a complete basis, but if Γ_i is degenerate ($d_i > 1$), then d_i orthogonal components will be required for each basis of Γ_i. It is then necessary to operate with P_i on d_i distinct elements of $\{\phi_n\}$ and to make the resulting functions orthogonal. Alternatively, a complete orthogonal basis may be generated directly from a single element, ϕ_n, by using the projection operators, $P_{i(kl)}$, which entail a knowledge of the irreducible matrix representations. The number of times, a_i, which an irreducible representation, Γ_i, is contained in that representation Γ spanned by $\{\phi_n\}$ cannot exceed the dimension, d_i of Γ_i: $a_i \leq d_i$. If $a_i < d_i$, then not all the bases generated by the d_i^2 $P_{i(kl)}$ operators will be independent. The choice of projection operators is made as indicated below:

Γ_i	d_i	a_i	$P_{i(kl)}$ operators to use
A or B	1	1	$P_{i(11)}$ $(= P_i)$
E	2	2	$(P_{i(11)}, P_{i(21)})$, $(P_{i(12)}, P_{i(22)})$
E	2	1	One of pairs above which does not result in zero when applied to ϕ_n.
F(T)	3	3	$(P_{i(11)}, P_{i(21)}, P_{i(31)})$, $(P_{i(12)}, P_{i(22)}, P_{i(32)})$ $(P_{i(13)}, P_{i(23)}, P_{i(33)})$
F(T)	3	2	Any two of the above triplets which yield linearly independent functions.
F(T)	3	1	One of the triplets above that yields a non-zero result.

3.7 Notation for irreducible representations (Mullikan notation)

Species Symbol	Character under E	Comments
A	+1	1-dimensional representation which is symmetric with respect to rotation about the principal axis.
B	+1	1-dimensional representation which is anti-symmetric with respect to rotation about the principal axis.
E	+2	2-dimensional representation.
F or T	+3	3-dimensional representation.
G or U	+4	4-dimensional representation.
H or V	+5	5-dimensional representation.

Subscripts 1, 2:

For A and B representations only - symmetric (1) or anti-symmetric (2) with respect to a C_2 axis perpendicular to the principal axis, or, in the absence of this, with respect to a vertical symmetry plane.

Primes and double primes, ' and ":

Where appropriate, these indicate symmetry (') or anti-symmetry (") with respect to a horizontal mirror plane.

g and u subscripts:

Where appropriate, these denote symmetry (g-gerade) or anti-symmetry (u-ungerade) with respect to an inversion centre.

If these rules allow several different labels, g and u take precedence over 1 and 2, which take precedence over ' and ". The labels of species of point groups $\mathbf{C}_{\infty v}$ and $\mathbf{D}_{\infty h}$ (linear molecules) are exceptional and are taken from the notation for the component of the electronic orbital angular momentum along the molecular axis.

3.8 Irreducible matrix representations

The two- and three-dimensional irreducible matrices for the generating operations of the point groups are given below. A complete irreducible matrix representation may be found by multiplying the generator matrices according to the relationships given for the symmetry operations of the point group (see also §3. 3). Non-degenerate (one-dimensional) and separably degenerate representations are, of course, contained explicitly in the character tables (§6).

Axial groups

As a standard basis for the generating operations we use the following:

A right-handed cartesian coordinate system;

The principal axis C_n or S_n lies in the z direction;

σ_v lies in the xz plane;

C'_2 lies in the x direction;

The sense of positive rotation as defined in §3. 1.

The axial point groups $\mathbf{C_n}$, $\mathbf{S_{2n}}$ and $\mathbf{C_{nh}}$ are cyclic and therefore Abelian; they are defined in terms of a single generator A ($A = C_n$ or S_n), and the relation $A^n = E$ which gives the n distinct symmetry operations:

$$E,\ A^k \quad (k = 1, 2, \ldots, n-1)\,.$$

The generator matrix for the kth separably degenerate irreducible representation, E_k, is:

$$E_k\colon A_k \rightarrow \begin{bmatrix} \epsilon^k & 0 \\ 0 & \epsilon^{k*} \end{bmatrix} \quad \text{where } \epsilon = e^{2\pi i/n} \quad \text{and} \begin{cases} k = 1, 2, \ldots \frac{1}{2}(n-1) & \text{if n odd} \\ k = 1, 2, \ldots \frac{1}{2}(n-2) & \text{if n even} \end{cases}$$

The axial point groups $\mathbf{C_{nv}}$, $\mathbf{D_n}$ and $\mathbf{D_{\frac{1}{2}nd}}$ are isomorphous and are defined in terms of two generators, A and B, and the relations. $A^n = B^2 = (AB)^2 = E$ which give the 2n distinct symmetry operations:

$$E,\ A^k,\ B,\ BA^k \quad (k = 1, 2, \ldots, n-1)\,.$$

The generator matrices for the kth doubly degenerate irreducible representation, E_k, are:

$$E_k\colon A_k \rightarrow \begin{bmatrix} \cos k\theta & -\sin k\theta \\ \sin k\theta & \cos k\theta \end{bmatrix} \quad \text{where } \theta = 2\pi/n \text{ and} \begin{cases} k = 1, 2, \ldots \frac{1}{2}(n-1) & \text{if n odd} \\ k = 1, 2, \ldots \frac{1}{2}(n-2) & \text{if n even} \end{cases}$$

$$E_k\colon B \rightarrow \begin{bmatrix} 1 & 0 \\ 0 & -1 \end{bmatrix}$$

where the generators A and B for the various groups are

	A	B
$\mathbf{C_{nv}}$	C_n	σ_v
$\mathbf{D_n}$	C_n	C'_2
$\mathbf{D_{\frac{1}{2}nd}}$	S_n	C'_2

The irreducible representations of the remaining axial groups, type $\mathbf{D_{nh}}$, may be obtained from those of $\mathbf{D_n}$ by a direct product, (see §3. 4).

Cubic groups

The standard basis uses the same coordinate system with the generators $C_2(\mathbf{T}, \mathbf{T_h})$, $C_4(\mathbf{O}, \mathbf{O_h})$ and $S_4(\mathbf{T_d})$ lying in the z direction. The special three-fold axis, C_3, lies in the positive quadrant and makes an angle of $\frac{\phi}{2}$ (54. 735... °) with the x, y and z directions.

Point group **T** is defined by the generators C_3 and C_2 (see §3. 3):

$$C_3^3 = C_2^2 = (C_2.C_3)^3 = E\,.$$

The following twelve distinct symmetry operations are generated:

$$E,\ C_3,\ C_3^2,\ C_3.C_2,\ C_3^2.C_2,\ C_2.C_3,\ C_2.C_3^2,\ C_2.C_3.C_2,\ C_2.C_3^2.C_2,\ C_2,\ C_3.C_2.C_3^2,\ C_3^2.C_2.C_3\,.$$

Matrices for the generators in the (separably degenerate) two-dimensional and three-dimensional irreducible representations, **E** and **F**, are:

$$\mathbf{E}: \begin{cases} C_3 \rightarrow \begin{bmatrix} \varepsilon & 0 \\ 0 & \varepsilon^* \end{bmatrix} & \varepsilon = e^{2\pi i/3} \\ C_2 \rightarrow \begin{bmatrix} 1 & 0 \\ 0 & 1 \end{bmatrix} \end{cases}$$

$$\mathbf{F}: \begin{cases} C_3 \rightarrow \begin{bmatrix} 0 & 0 & 1 \\ 1 & 0 & 0 \\ 0 & 1 & 0 \end{bmatrix} \\ C_2 \rightarrow \begin{bmatrix} -1 & 0 & 0 \\ 0 & -1 & 0 \\ 0 & 0 & 1 \end{bmatrix} \end{cases}$$

The irreducible matrix representations of $\mathbf{T_h}$ are obtained by direct product: $\mathbf{T_h} = \mathbf{C_i} \times \mathbf{T}$.

Point groups **O** and $\mathbf{T_d}$ are isomorphous. The generating relations for **O** are:

$$C_3^3 = C_4^4 = (C_4 . C_3)^2 = E$$

giving the twenty-four distinct symmetry operations:

$$E,\ C_3,\ C_3^2,\ C_4,\ C_4^2,\ C_4^3,\ C_3.C_4,\ C_3.C_4^2,\ C_3.C_4^3,\ C_3^2.C_4,\ C_3^2.C_4^2,\ C_3^2.C_4^3,\ C_4^2.C_3,$$

$$C_4^3.C_3,\ C_4.C_3^2,\ C_4^2.C_3^2,\ C_3.C_4^2.C_3,\ C_3.C_4^3.C_3,\ C_3.C_4^2.C_3^2,\ C_3^2.C_4^2.C_3^2,\ C_3^2.C_4^2.C_3,$$

$$C_4.C_3^2.C_4^2,\ C_4.C_3^2.C_4^2.C_3,\ C_4.C_3^2.C_4^2.C_3^2 .$$

Matrices for the generators in the **E**, $\mathbf{F_1}$ and $\mathbf{F_2}$ irreducible representations are:

$$\mathbf{E}: \begin{cases} C_3^{xyz} \rightarrow \begin{bmatrix} -\frac{1}{2} & -\frac{1}{2}\sqrt{3} \\ \frac{1}{2}\sqrt{3} & -\frac{1}{2} \end{bmatrix} \\ C_4(z) \rightarrow \begin{bmatrix} 1 & 0 \\ 0 & -1 \end{bmatrix} \end{cases}$$

$$\mathbf{F_1}: \begin{cases} C_3^{xyz} \rightarrow \begin{bmatrix} 0 & 0 & 1 \\ 1 & 0 & 0 \\ 0 & 1 & 0 \end{bmatrix} \\ C_4(z) \rightarrow \begin{bmatrix} 0 & -1 & 0 \\ 1 & 0 & 0 \\ 0 & 0 & 1 \end{bmatrix} \end{cases}$$

$$\mathbf{F_2}: \begin{cases} C_3^{xyz} \rightarrow \begin{bmatrix} 0 & 0 & 1 \\ 1 & 0 & 0 \\ 0 & 1 & 0 \end{bmatrix} \\ C_4(z) \rightarrow \begin{bmatrix} 0 & 1 & 0 \\ -1 & 0 & 0 \\ 0 & 0 & -1 \end{bmatrix} \end{cases}$$

$\mathbf{T_d}$ is isomorphous with **O**, on substitution of the generator S_4^3 for C_4. $\mathbf{O_h}$ is obtained by the direct product:

$$\mathbf{O_h} = \mathbf{C_i} \times \mathbf{O}.$$

4 Crystallographic aspects of point symmetry

4.1 Relation between the Schönflies and the Hermann-Mauguin (international) notations

Symmetry operations: The proper rotation is a basic operation common to both systems, but the improper operation in the international system is defined as a rotation-**inversion**, viz. rotation about the axis through $\frac{2\pi k}{n}$ radians combined with inversion through the origin. Such a rotation-inversion, $\bar{n}_k$, is equivalent to a rotation-reflection through an angle $\pi + \frac{2\pi k}{n}$. Thus $\bar{n}_k = C_n^k . i$.

Proper rotations		Improper rotations	
Schönflies	**International**	**Schönflies**	**International**
$C_1^1 = E$	1	$S_1^1 = \sigma$	$\bar{2} = m$
$C_2^1 = C_2$	2	$S_2^1 = i$	$\bar{1}$
$C_3^1 = C_3$	3	$S_3^1 = S_3$	$\bar{6}_5$
C_3^2	3_2	S_3^5	$\bar{6}$
$C_4^1 = C_4$	4	$S_4^1 = S_4$	$\bar{4}_3$
C_4^3	4_3	S_4^3	$\bar{4}$
$C_6^1 = C_6$	6	$S_6^1 = S_6$	$\bar{3}_5$
C_6^5	6_5	S_6^5	$\bar{3}$
$C_n^{(k)}$	$n_{(k)}$		

If $k = 1$, the suffix is customarily omitted.

4.2 The crystal systems and crystallographic point groups

System	Schönflies symbol	International symbol	Number of space group	Order	System properties
Triclinic	C_1	1	1	1	$a \neq b \neq c$; $\alpha \neq \beta \neq \gamma \neq \frac{\pi}{2}$
	$C_i(S_2)$	$\bar{1}$	2	2	P [2]
Mono-clinic	C_2	2	3-5	2	$a \neq b \neq c$; $\alpha = \gamma = \frac{\pi}{2} \neq \beta$
	$C_s(C_{1h})$	m	6-9	2	P, I [6]
	C_{2h}	$\frac{2}{m}$	10-15	4	
Ortho-rhombic	D_2	222	16-24	4	$a \neq b \neq c$; $\alpha = \beta = \gamma = \frac{\pi}{2}$
	C_{2v}	mm2 (mm)	25-46	4	P, I, F, C [12]
	D_{2h}	$\frac{2}{m}\frac{2}{m}\frac{2}{m}$ (mmm)	47-74	8	
Tetra-gonal	C_4	4	75-80	4	$a = b \neq c$;
	S_4	$\bar{4}$	81, 82	4	$\alpha = \beta = \gamma = \frac{\pi}{2}$
	C_{4h}	$\frac{4}{m}$	83-88	8	P, I [14]

System	Schönflies symbol	International symbol	Number of space group	Order	System properties
	D_4	422	89-98	8	
	C_{4v}	4mm	99-110	8	
	$D_{2d}(V_d)$	$\bar{4}2m$	111-122	8	
	D_{4h}	$\frac{4}{m}\frac{4}{m}\frac{4}{m}$ $(\frac{4}{mmm})$	123-142	16	
Trigonal (may be taken as a sub-division of hex-agonal)	C_3	3	143-146	3	Rhombohedral axes:
	S_6	$\bar{3}$	147-148	6	$a = b = c$;
	D_3	32	149-155	6	$\alpha = \beta = \gamma < \frac{2\pi}{3} \neq \frac{\pi}{2}$ or $\frac{\pi}{3}$. R
	C_{3v}	3m	156-161	6	Hexagonal axes:
	D_{3d}	$\bar{3}\frac{2}{m}$ $(\bar{3}m)$	162-167	12	$a = b \neq c$; $\alpha = \beta = \frac{\pi}{2}$; $\gamma = \frac{2\pi}{3}$
					P [10]
Hexagonal	C_6	6	168-173	6	$a = b \neq c$;
	C_{3h}	$\frac{3}{m} = \bar{6}$	174	6	$\alpha = \beta = \frac{\pi}{2}$; $\gamma = \frac{2\pi}{3}$.
	C_{6h}	$\frac{6}{m}$	175, 176	12	P [7]
	D_6	622	177-182	12	
	C_{6v}	6mm	183-186	12	
	D_{3h}	$\frac{3}{m}2 = \bar{6}m2$	187-190	12	
	D_{6h}	$\frac{6}{m}\frac{2}{m}\frac{2}{m}$ $(\frac{6}{mmm})$	191-194	24	
Cubic	T	23	195-199	12	$a = b = c$;
	T_h	$\frac{2}{m}\bar{3}$ (m3)	200-206	24	$\alpha = \beta = \gamma = \frac{\pi}{2}$
	O	432	207-214	24	P, I, F [15]
	T_d	$\bar{4}3m$	215-220	24	
	O_h	$\frac{4}{m}\bar{3}\frac{2}{m}$ (m3m)	221-230	48	

Acceptable short symbols are given in parentheses. The system properties listed are: the unit cell geometry, the number of Bravais space groups in brackets [ ] and the Bravais unit cell types,

P primitive
I body-centred
C end-centred
F face-centred
R primitive rhombohedral cell .

For a list of the crystallographic space groups, their symbols and site symmetries, see §7. 3.

4. 3 The point groups in stereographic projection

In the stereographic projections, the following conventions have been adopted:

A proper axis, C_n, is denoted by an open (unshaded) polygon of appropriate order.

An improper (rotation-reflection) axis, S_n, is only marked when its order exceeds (by a factor of 2) that of the coincident proper axis; it is denoted by two crossed polygons, alternately shaded and open.

Mirror planes are denoted by solid lines (including the equatorial plane).

Other axes and the outline of the stereogram are dashed lines, unless they coincide with mirror planes.

A general point and all its symmetrically equivalent points are represented by crosses (upper hemisphere) and circles (lower hemisphere).

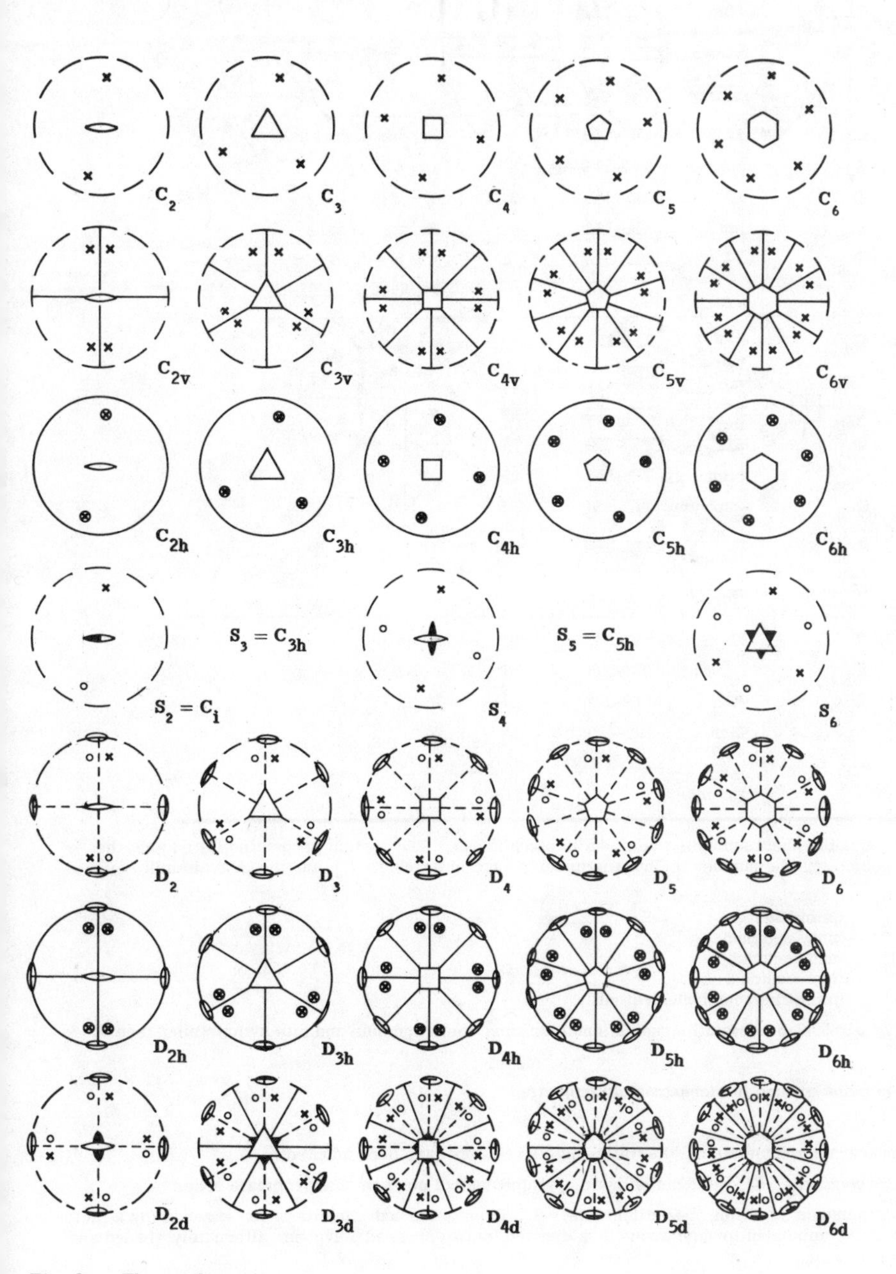

Fig. 1. The axial groups.

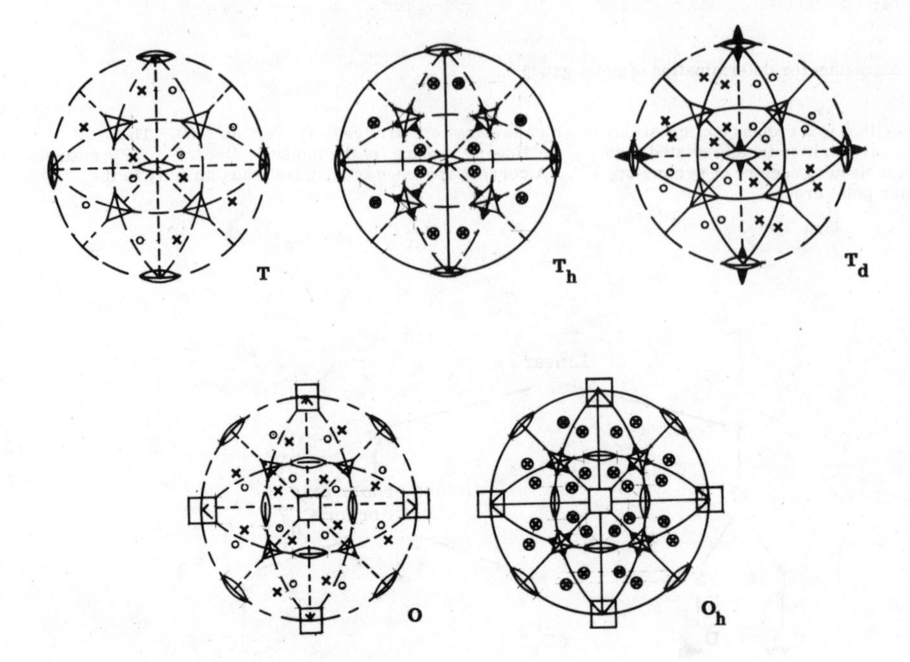

Fig. 2. The cubic groups.

5 Some aids to the use of character tables

5.1 Algorithm for determination of point group

The algorithm is a sequence of questions to which the answers are 'yes' (Y) or 'no' (N). The questions are written in abbreviated form, e.g. 'linear?' means 'is the molecule linear?': 'i?' means 'has the molecule a centre of symmetry, i?'. A correct sequence of answers leads directly to the molecular point group.

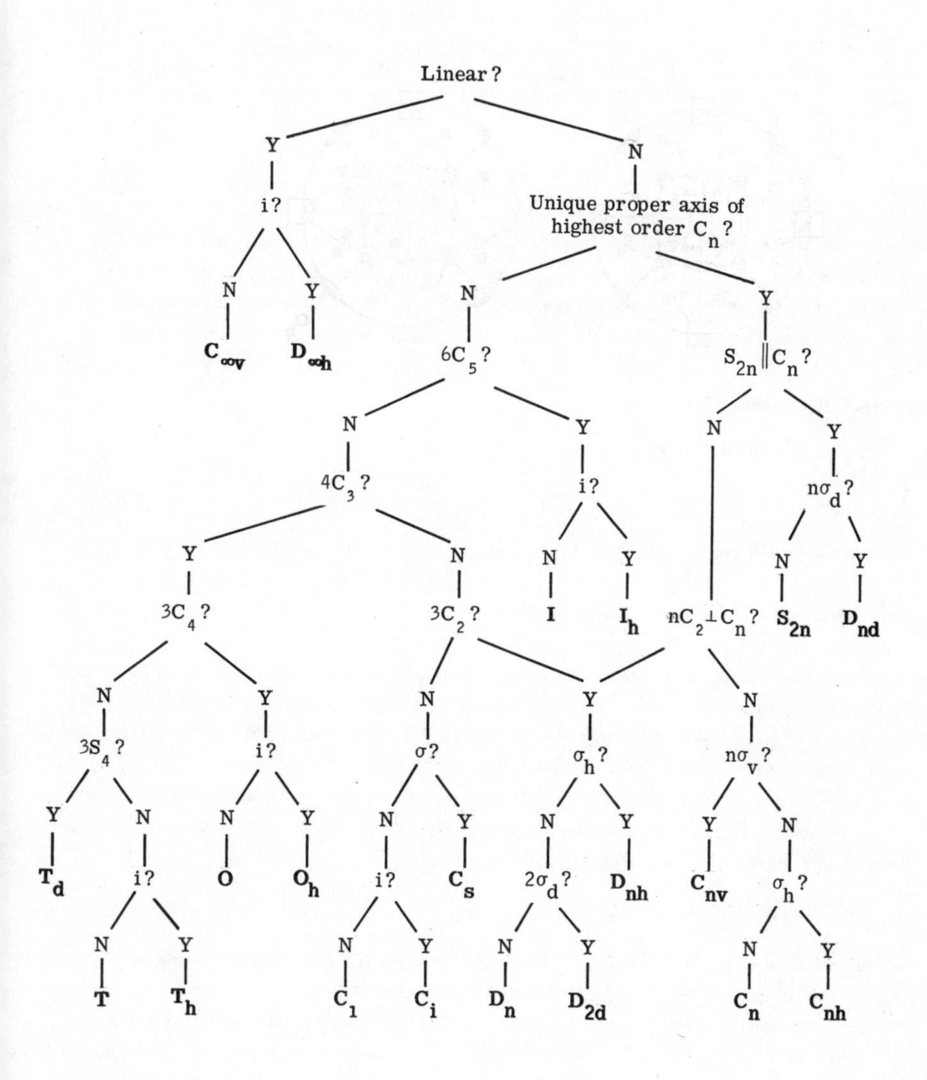

5.2 Formulae for reduction of representations

a_i = number of times that the ith irreducible representation, Γ_i, occurs in the reducible representation, Γ: $\Gamma = \sum_i a_i \Gamma_i$.

$\chi(R)$ = character of the reducible representation of Γ under an operation of the Rth class.

$\chi_i(R)$ = character of the ith irreducible representation Γ_i under an operation of the Rth class.

g_R = number of operations in the Rth class.

h = order of the group.

For finite groups:

$$a_i = \frac{1}{h} \sum_R g_R . \chi(R) . \chi_i(R) \qquad \text{the sum goes over all classes of the group.}$$

For infinite (continuous) groups[6] ($\mathbf{C}_{\infty v}$ and $\mathbf{D}_{\infty h}$):

$$a_i = \frac{\sum_R \int_0^{2\pi} \chi(R) . \chi_i(R) . d\phi}{\sum_R \int_0^{2\pi} d\phi}$$

The sum goes over the classes of distinct operations only (i. e. the classes of C_∞^ϕ and σ_v in $\mathbf{C}_{\infty v}$, and the classes of C_∞^ϕ, σ_v, S_∞^ϕ and C_2' in $\mathbf{D}_{\infty h}$; other classes of operation (E, i), sometimes included in the character tables, are merely special cases of C_∞^ϕ and S_∞^ϕ. Within each class the integration is performed over the range of the continuous variable, ϕ, denoting the rotation angle; i. e. from 0 to 2π.

The following relationships are useful:

$$\int_0^{2\pi} \cos k\phi . \cos l\phi . d\phi = \pi \delta_{kl}$$

where δ_{kl} is the Kronecker delta, $\delta_{kl} = 1$ for $k = l$, $\delta_{kl} = 0$ for $k \neq l$;

and $\int_0^{2\pi} \cos k\phi . d\phi = 0$ for k integer .

5.3 The representation for normal modes of vibration

In an N-atomic molecule the representation for generalised nuclear displacements is set up in terms of 3N Cartesian displacement coordinates; if the ith atom is unshifted (i. e. sent into itself) under a symmetry operation, its cartesian displacement coordinates, Δx_i, Δy_i, Δz_i, will contribute an amount $\pm 1 + 2\cos\theta$ to the character of the representation, where + refers to a C_n^k and - to an S_n^k operation, and $\theta = \frac{2\pi k}{n}$. If an atom is shifted (i. e. permuted with its equivalent atoms) its displacement coordinates contribute nothing to the character. Accordingly, the character of the reducible representation under an operation of the Rth class $\chi_{cart}(R)$ is given by:

$$\chi_{cart}(R) = n_R(\pm 1 + 2\cos\theta)$$

where n_R is the number of atoms unshifted by an operation of the Rth class and the values of $(\pm 1 + 2\cos\theta)$ for C_n^k and S_n^k respectively may be found in table 5. 4.

The representations so obtained will contain, in addition to the internal vibrational degrees of freedom, the pure translations (x, y, z) and rotations (R_x, R_y, R_z) of the molecule. These must be subtracted from the representation either before (as character sums) or after (by consultation of the table) it is reduced, to yield the 3N - 6 degrees of vibrational freedom (3N - 5 for linear molecules). Alternatively,[7] the translations (character $\pm 1 + 2\cos\theta$) and rotations (character $1 \pm 2\cos\theta$) may be eliminated at the outset by subtracting their total contribution to character:

$$\chi_{vib}(C_n^k) = (n_R - 2)(1 + 2\cos\theta) ,$$
$$\chi_{vib}(S_n^k) = n_R(-1 + 2\cos\theta) .$$

$\chi_{vib}(R)$ is then the character of the representation spanned by the normal modes of vibration only.

5.4 Rotation angles and their cosines

Rotation angles $\theta = \frac{2\pi k}{n}$		cos θ		Operations and contribution to character			
				Proper		Improper	
Radians	Degrees			C_n^k	$+1+2\cos\theta$	S_n^k	$-1+2\cos\theta$
$0, 2\pi$	0, 360	+1	+1.0000	$C_1^1 = E$	+3	$S_1^1 = \sigma$	+1
π	180	-1	-1.0000	C_2^1	-1	$S_2^1 = i$	-3
$\frac{2\pi}{3}, \frac{4\pi}{3}$	120, 240	$-\frac{1}{2}$	-0.5000	C_3^1, C_3^2	0	S_3^1, S_3^5	-2
$\frac{\pi}{2}, \frac{3\pi}{2}$	90, 270	0	0.0000	C_4^1, C_4^3	+1	S_4^1, S_4^3	-1
$\frac{2\pi}{5}, \frac{8\pi}{5}$	72, 288	$\frac{1}{2}(\tau-1) = \frac{1}{4}(\sqrt{5}-1)$	+0.3090	C_5^1, C_5^4	$+\tau$	S_5^1, S_5^9	$\tau-2$
$\frac{4\pi}{5}, \frac{6\pi}{5}$	144, 216	$-\frac{\tau}{2} = -\frac{1}{4}(\sqrt{5}+1)$	-0.8090	C_5^2, C_5^3	$1-\tau$	S_5^3, S_5^7	$-1-\tau$
$\frac{\pi}{3}, \frac{5\pi}{3}$	60, 300	$+\frac{1}{2}$	+0.5000	C_6^1, C_6^5	+2	S_6^1, S_6^5	0
$\frac{2\pi}{7}, \frac{12\pi}{7}$	51.43, 308.57		+0.6234	C_7^1, C_7^6	$+1+2\cos\frac{2\pi}{7}$	S_7^1, S_7^{13}	$-1+\cos\frac{2\pi}{7}$
$\frac{4\pi}{7}, \frac{10\pi}{7}$	102.86, 257.14		-0.2227	C_7^2, C_7^5	$+1+2\cos\frac{4\pi}{7}$	S_7^3, S_7^{11}	$-1+2\cos\frac{4\pi}{7}$
$\frac{6\pi}{7}, \frac{8\pi}{7}$	154.28, 205.72		-0.9010	C_7^3, C_7^4	$+1+2\cos\frac{6\pi}{7}$	S_7^5, S_7^9	$-1+2\cos\frac{6\pi}{7}$
$\frac{\pi}{4}, \frac{7\pi}{4}$	45, 315	$\frac{+1}{\sqrt{2}}$	+0.7071	C_8^1, C_8^7	$1+\sqrt{2}$	S_8^1, S_8^7	$-1+\sqrt{2}$
$\frac{3\pi}{4}, \frac{5\pi}{4}$	135, 225	$\frac{-1}{\sqrt{2}}$	-0.7071	C_8^3, C_8^5	$1-\sqrt{2}$	S_8^3, S_8^5	$-1-\sqrt{2}$
$\frac{\pi}{5}, \frac{9\pi}{5}$	36, 324	$\frac{\tau}{2} = \frac{1}{4}(\sqrt{5}+1)$	+0.8090	C_{10}^1, C_{10}^9	$1+\tau$	S_{10}^1, S_{10}^9	$-1+\tau$
$\frac{3\pi}{5}, \frac{7\pi}{5}$	108, 252	$\frac{1}{2}(1-\tau) = \frac{1}{4}(1-\sqrt{5})$	-0.3090	C_{10}^3, C_{10}^7	$2-\tau$	S_{10}^3, S_{10}^7	$-\tau$
$\frac{\pi}{6}, \frac{11\pi}{6}$	30, 330	$\frac{\sqrt{3}}{2}$	+0.8660	C_{12}^1, C_{12}^{11}	$1+\sqrt{3}$	S_{12}^1, S_{12}^{11}	$-1+\sqrt{3}$
$\frac{5\pi}{6}, \frac{7\pi}{6}$	150, 210	$-\frac{\sqrt{3}}{2}$	-0.8660	C_{12}^5, C_{12}^7	$1-\sqrt{3}$	S_{12}^5, S_{12}^7	$-1-\sqrt{3}$
$\frac{\pi}{7}, \frac{13\pi}{7}$	25.71, 334.29		+0.8853	C_{14}^1, C_{14}^{13}	$+1+2\cos\frac{\pi}{7}$	S_{14}^1, S_{14}^{13}	$-1+2\cos\frac{\pi}{7}$
$\frac{3\pi}{7}, \frac{11\pi}{7}$	77.13, 282.87		+0.2227	C_{14}^3, C_{14}^{11}	$+1+2\cos\frac{3\pi}{7}$	S_{14}^3, S_{14}^{11}	$-1+2\cos\frac{3\pi}{7}$
$\frac{5\pi}{7}, \frac{9\pi}{7}$	128.55, 231.45		-0.6234	C_{14}^5, C_{14}^9	$+1+\cos\frac{5\pi}{7}$	S_{14}^5, S_{14}^9	$-1+2\cos\frac{5\pi}{7}$
$\frac{\pi}{8}, \frac{15\pi}{8}$	22.5, 337.5	$\frac{1}{2}(2+\sqrt{2})^{\frac{1}{2}}$	+0.9239	C_{16}^1, C_{16}^{15}	$1+(2+\sqrt{2})^{\frac{1}{2}}$	S_{16}^1, S_{16}^{15}	$-1+(2+\sqrt{2})^{\frac{1}{2}}$
$\frac{3\pi}{8}, \frac{13\pi}{8}$	67.5, 192.5	$\frac{1}{2}(2-\sqrt{2})^{\frac{1}{2}}$	+0.3827	C_{16}^3, C_{16}^{13}	$1+(2-\sqrt{2})^{\frac{1}{2}}$	S_{16}^3, S_{16}^{13}	$-1+(2-\sqrt{2})^{\frac{1}{2}}$
$\frac{5\pi}{8}, \frac{11\pi}{8}$	112.5, 247.5	$-\frac{1}{2}(2-\sqrt{2})^{\frac{1}{2}}$	-0.3827	C_{16}^5, C_{16}^{11}	$1-(2-\sqrt{2})^{\frac{1}{2}}$	S_{16}^5, S_{16}^{11}	$-1-(2-\sqrt{2})^{\frac{1}{2}}$
$\frac{7\pi}{8}, \frac{9\pi}{8}$	157.5, 202.5	$-\frac{1}{2}(2+\sqrt{2})^{\frac{1}{2}}$	-0.9239	C_{16}^7, C_{16}^9	$1-(2+\sqrt{2})^{\frac{1}{2}}$	S_{16}^7, S_{16}^9	$-1-(2+\sqrt{2})^{\frac{1}{2}}$

5.5 Transformations of functions

In the **axial groups**, certain of the linear, quadratic and cubic orthogonal functions may frequently occur alone, or as partner functions in doubly degenerate species. If so, the following formulae will yield the characters, χ_f, of their irreducible representations directly. If the paired functions are separable however, the representation given by the formulae will be reducible and the transformations of the individual functions should then be examined.

Functions	$\chi_f(C_n^k(z))$ [+ sign] $\chi_f(S_n^k(z))$ [- sign]	$\chi_f(C_2')$ $\chi_f(C_2'')$	$\chi_f(\sigma_v)$ $\chi_f(\sigma_d)$
z, z^3, $z(x^2+y^2)$	± 1	-1	$+1$
z^2, x^2+y^2, R_z	$+1$	$+1$	$+1$
[x, y], [xz^2, yz^2], [$x(x^2+y^2)$, $y(x^2+y^2)$]	$+2\cos\theta$	0	0
[xz, yz], [R_x, R_y]	$\pm 2\cos\theta$	0	0
[x^2-y^2, xy]	$+2\cos 2\theta$	0	0
[xyz, $z(x^2-y^2)$]	$\pm 2\cos 2\theta$	0	0
[$x(x^2-3y^2)$, $y(3x^2-y^2)$]	$+2\cos 3\theta$	0	0

$$\left(\theta = \frac{2\pi k}{n}\right)$$

In **axial** groups of 2- or 4-fold symmetry the basis set of cubic functions takes a simpler form.[8] The conventional basis sets of 1-electron orbitals transform as the following functions:

p orbitals x, y, z

d orbitals $2z^2 - x^2 - y^2$, $x^2 - y^2$, xy, xz, yz

f orbitals[9] $z(5z^2 - 3r^2)$, $x(5z^2 - 3r^2)$, $y(5z^2 - 3r^2)$, xyz, $z(x^2 - y^2)$, $x(x^2 - 3y^2)$, $y(3x^2 - y^2)$.

In the **cubic** groups, other basis sets may be more conveniently chosen.

6 Point group character tables

6.1 The non-axial groups

C_1	E (h = 1)
A	+1

C_i	E	i	(h = 2)		
A_g	+1	+1	R_x, R_y, R_z	$\{x^2, y^2, z^2$ / xy, xz, yz	
A_u	+1	-1	x, y, z		all cubic functions

C_s	E	σ_h	(h = 2)		
A'	+1	+1	x, y, R_z	$\{x^2, y^2$ / z^2, xy	$xz^2, yz^2, x^2y, xy^2, x^3, y^3$
A"	+1	-1	z, R_x, R_y	yz, xz	z^3, xyz, y^2z, x^2z

6.2 The axial groups

The cyclic groups

C_2	E	$C_2(z)$	(h = 2)		
A	+1	+1	z, R_z	x^2, y^2, z^2, xy	z^3, xyz, y^2z, x^2z
B	+1	-1	x, y, R_x, R_y	yz, xz	$xz^2, yz^2, x^2y, xy^2, x^3, y^3$

C_3	E	$C_3(z)$	C_3^2	(h = 3); $\varepsilon=\exp(2\pi i/3)$		
A	+1	+1	+1	z, R_z	x^2+y^2, z^2	$z^3, y(3x^2-y^2), x(x^2-3y^2), z(x^2+y^2)$
E	+1	ε	ε^*	$x+iy; R_x+iR_y$	(x^2-y^2, xy)	(xz^2, yz^2) $[xyz, z(x^2-y^2)]$ $[x(x^2+y^2), y(x^2+y^2)]$
	+1	ε^*	ε	$x-iy; R_x-iR_y$	(yz, xz)	

C_4	E	$C_4(z)$	C_2	C_4^3	(h = 4)		
A	+1	+1	+1	+1	z, R_z	x^2+y^2, z^2	$z^3, z(x^2+y^2)$
B	+1	-1	+1	-1		x^2-y^2, xy	$xyz, z(x^2-y^2)$
E	+1	+i	-1	-i	$x+iy; R_x+iR_y$	(yz, xz)	$(xz^2, yz^2)(xy^2, x^2y)(x^3, y^3)$
	+1	-i	-1	+i	$x-iy; R_x-iR_y$		

C_5	E	C_5	C_5^2	C_5^3	C_5^4	(h = 5); $\varepsilon=\exp(2\pi i/5)$		
A	+1	+1	+1	+1	+1	z, R_z	x^2+y^2, z^2	$z^3, z(x^2+y^2)$
E_1	+1	ε	ε^2	ε^{2*}	ε^*	$x+iy; R_x+iR_y$	(yz, xz)	$(xz^2, yz^2)[x(x^2+y^2), y(x^2+y^2)]$
	+1	ε^*	ε^{2*}	ε^*	ε	$x-iy; R_x-iR_y$		
E_2	+1	ε^2	ε^*	ε	ε^{2*}		(x^2-y^2, xy)	$[xyz, z(x^2-y^2)][y(3x^2-y^2), x(x^2-3y^2)]$
	+1	ε^{2*}	ε	ε^*	ε^2			

C_6	E	C_6	C_3	C_2	C_3^2	C_6^5	(h = 6); $\varepsilon = \exp(2\pi i/6)$		
A	+1	+1	+1	+1	+1	+1	z, R_z	x^2+y^2, z^2	z^3, $z(x^2+y^2)$
B	+1	-1	+1	-1	+1	-1			$y(3x^2-y^2)$, $x(x^2-3y^2)$
E_1	+1	ε	$-\varepsilon^*$	-1	$-\varepsilon$	ε^*	x+iy; R_z+iR_y	(xz, yz)	(xz^2, yz^2) $[x(x^2+y^2), y(x^2+y^2)]$
	+1	ε^*	$-\varepsilon$	-1	$-\varepsilon^*$	ε	x-iy; R_x-iR_y		
E_2	+1	$-\varepsilon^*$	$-\varepsilon$	+1	$-\varepsilon^*$	$-\varepsilon$		(x^2-y^2, xy)	$[xyz, z(x^2-y^2)]$
	+1	$-\varepsilon$	$-\varepsilon^*$	+1	$-\varepsilon$	$-\varepsilon^*$			

C_7	E	C_7	C_7^2	C_7^3	C_7^4	C_7^5	C_7^6	(h = 7); $\varepsilon = \exp(2\pi i/7)$		
A	+1	+1	+1	+1	+1	+1	+1	z, R_z	x^2+y^2, z^2	z^3, $z(x^2+y^2)$
E_1	+1	ε	ε^2	ε^3	ε^{3*}	ε^{2*}	ε^*	x+iy; R_x+iR_y	(xz, yz)	$(xz^2, yz^2)$$[x(x^2+y^2), y(x^2+y^2)]$
	+1	ε^*	ε^{2*}	ε^{3*}	ε^3	ε^2	ε	x-iy; R_x-iR_y		
E_2	+1	ε^2	ε^{3*}	ε^*	ε	ε^3	ε^{2*}		(x^2-y^2, xy)	$[xyz, z(x^2-y^2)]$
	+1	ε^{2*}	ε^3	ε	ε^*	ε^{3*}	ε^2			
E_3	+1	ε^3	ε^*	ε^2	ε^{2*}	ε	ε^{3*}			$[y(3x^2-y^2), x(x^2-3y^2)]$
	+1	ε^{3*}	ε	ε^{2*}	ε^2	ε^*	ε^3			

C_8	E	C_8	C_4	C_8^3	C_2	C_8^5	C_4^3	C_8^7	(h = 8); $\varepsilon = \exp(2\pi i/8)$		
A	+1	+1	+1	+1	+1	+1	+1	+1	z, R_z	x^2+y^2, z^2	$z^3, z(x^2+y^2)$
B	+1	-1	+1	-1	+1	-1	+1	-1			
E_1	+1	ε	i	$-\varepsilon^*$	-1	$-\varepsilon$	-i	ε^*	$x+iy; R_x+iR_y$	(xz, yz)	(xz^2, yz^2) $[x(x^2+y^2), y(x^2+y^2)]$
	+1	ε^*	-i	$-\varepsilon$	-1	$-\varepsilon^*$	i	ε	$x-iy; R_x-iR_y$		
E_2	+1	i	-1	-i	+1	i	-1	-i		(x^2-y^2, xy)	$[xyz, z(x^2-y^2)]$
	+1	-i	-1	i	+1	-i	-1	i			
E_3	+1	$-\varepsilon$	i	ε^*	-1	ε	-i	$-\varepsilon^*$			$[y(3x^2-y^2), x(x^2-3y^2)]$
	+1	$-\varepsilon^*$	-i	ε	-1	ε^*	i	$-\varepsilon$			

S_4	E	S_4	$C_2(z)$	S_4^3	(h = 4)		
A	+1	+1	+1	+1	R_z	x^2+y^2, z^2	$xyz, z(x^2-y^2)$
B	+1	-1	+1	-1	z	x^2-y^2, xy	$z^3, z(x^2+y^2)$
E	+1	+i	-1	-i	$x+iy; R_x+iR_y$	(xz, yz)	(xz^2, yz^2) (xy^2, x^2y) (x^3, y^3)
	+1	-i	-1	+i	$x-iy; R_x-iR_y$		

S_6	E	$C_3(z)$	C_3^2	i	S_6^5	S_6	(h = 6); $\varepsilon = \exp(2\pi i/3)$		
A_g	+1	+1	+1	+1	+1	+1	R_z	x^2+y^2, z^2	
E_g	+1	ε	ε^*	+1	ε	ε^*	R_x+iR_y	(x^2-y^2, xy) (xz, yz)	
	+1	ε^*	ε	+1	ε^*	ε	R_x-iR_y		
A_u	+1	+1	+1	-1	-1	-1	z		$z^3, y(3x^2-y^2), x(x^2-3y^2), z(x^2+y^2)$
E_u	+1	ε	ε^*	-1	$-\varepsilon$	$-\varepsilon^*$	x+iy		(xz^2, yz^2) $[xyz, z(x^2-y^2)]$ $[x(x^2+y^2). y(x^2+y^2)]$
	+1	ε^*	ε	-1	$-\varepsilon^*$	$-\varepsilon$	x-iy		

$\mathbf{S_8}$	E	S_8	$C_4(z)$	S_8^3	C_2	S_8^5	C_4^3	S_8^7	(h = 8); $\varepsilon = \exp(2\pi i/8)$		
A	+1	+1	+1	+1	+1	+1	+1	+1	R_z	$x^2+y^2,\ z^2$	
B	+1	-1	+1	-1	+1	-1	+1	-1	z		$z^3,\ z(x^2+y^2)$
E_1	+1	ε	+i	$-\varepsilon^*$	-1	$-\varepsilon$	-i	ε^*	x+iy		$(xz^2,\ yz^2)$ $[x(x^2+y^2),\ y(x^2+y^2)]$
	+1	ε^*	-i	$-\varepsilon$	-1	$-\varepsilon^*$	+i	ε	x-iy		
E_2	+1	+i	-1	-i	+1	+i	-1	-i		$(x^2-y^2,\ xy)$	$[xyz,\ z(x^2-y^2)]$
	+1	-i	-1	+i	+1	-i	-1	+i			
E_3	+1	$-\varepsilon$	+i	ε^*	-1	ε	-i	$-\varepsilon^*$	R_x+iR_y	$(xz,\ yz)$	$[y(3x^2-y^2),\ x(x^2-3y^2)]$
	+1	$-\varepsilon^*$	-i	ε	-1	ε^*	+i	$-\varepsilon$	R_x-iR_y		

$\mathbf{S_{10}}$	E	C_5	C_5^2	C_5^3	C_5^4	i	S_{10}^7	S_{10}^9	S_{10}	S_{10}^3	(h = 10); $\varepsilon = \exp(2\pi i/5)$		
A_g	+1	+1	+1	+1	+1	+1	+1	+1	+1	+1	R_z	$z^2,\ x^2+y^2$	
E_{1g}	+1	ε	ε^2	ε^{2*}	ε^*	+1	ε	ε^2	ε^{2*}	ε^*	R_x+iR_y	$(xz,\ yz)$	
	+1	ε^*	ε^{2*}	ε^2	ε	+1	ε^*	ε^{2*}	ε^2	ε	R_x-iR_y		
E_{2g}	+1	ε^2	ε^*	ε	ε^{2*}	+1	ε^2	ε^*	ε	ε^{2*}		$(x^2-y^2,\ xy)$	
	+1	ε^{2*}	ε	ε^*	ε^2	+1	ε^{2*}	ε	ε^*	ε^2			
B_u	+1	+1	+1	+1	+1	-1	-1	-1	-1	-1	z		$z^3,\ z(x^2+y^2)$
E_{1u}	+1	ε	ε^2	ε^{2*}	ε^*	-1	$-\varepsilon$	$-\varepsilon^2$	$-\varepsilon^{2*}$	$-\varepsilon^*$	x+iy		$(xz^2,\ yz^2)$ $[x(x^2+y^2),\ y(x^2+y^2)]$
	+1	ε^*	ε^{2*}	ε^2	ε	-1	$-\varepsilon^*$	$-\varepsilon^{2*}$	$-\varepsilon^2$	$-\varepsilon$	x-iy		
E_{2u}	+1	ε^2	ε^*	ε	ε^{2*}	-1	$-\varepsilon^2$	$-\varepsilon^*$	$-\varepsilon$	$-\varepsilon^{2*}$			$[xyz,\ z(x^2-y^2)]$ $[y(3x^2-y^2),\ x(x^2-3y^2)]$
	+1	ε^{2*}	ε	ε^*	ε^2	-1	$-\varepsilon^{2*}$	$-\varepsilon$	$-\varepsilon^*$	$-\varepsilon^2$			

S_{12}	E	S_{12}	C_6	S_4	C_3	S_{12}^5	C_2	S_{12}^7	C_3^2	S_4^3	C_6^5	S_{12}^{11}	(h = 12);	$\varepsilon = \exp(2\pi i/12)$	
A	+1	+1	+1	+1	+1	+1	+1	+1	+1	+1	+1	+1	R_z	z^2, x^2+y^2	
B	+1	-1	+1	-1	+1	-1	+1	-1	+1	-1	+1	-1	z		z^3, $z(x^2+y^2)$
E_1	+1	ε	ε^2	+i	$-\varepsilon^{2*}$	$-\varepsilon^*$	-1	$-\varepsilon$	$-\varepsilon^2$	-i	ε^{2*}	ε^*	x+iy, x-iy		(xz^2, yz^2) $[x(x^2+y^2), y(x^2+y^2)]$
	+1	ε^*	ε^{2*}	-i	$-\varepsilon^2$	$-\varepsilon$	-1	$-\varepsilon^*$	$-\varepsilon^{2*}$	+i	ε^2	ε			
E_2	+1	ε^2	$-\varepsilon^{2*}$	-1	$-\varepsilon^2$	ε^{2*}	+1	ε^2	$-\varepsilon^{2*}$	-1	$-\varepsilon^2$	ε^{2*}		(x^2-y^2, xy)	
	+1	ε^{2*}	$-\varepsilon^2$	-1	$-\varepsilon^{2*}$	ε^2	+1	ε^{2*}	$-\varepsilon^2$	-1	$-\varepsilon^{2*}$	ε^2			
E_3	+1	+i	-1	-i	+1	+i	-1	-i	+1	+i	-1	-i			$[x(x^2-3y^2), y(3x^2-y^2)]$
	+1	-i	-1	+i	+1	-i	-1	+i	+1	-i	-1	+i			
E_4	+1	$-\varepsilon^{2*}$	$-\varepsilon^2$	+1	$-\varepsilon^{2*}$	$-\varepsilon^2$	+1	$-\varepsilon^{2*}$	$-\varepsilon^2$	+1	$-\varepsilon^{2*}$	$-\varepsilon^2$			$[xyz, z(x^2-y^2)]$
	+1	$-\varepsilon^2$	$-\varepsilon^{2*}$	+1	$-\varepsilon^2$	$-\varepsilon^{2*}$	+1	$-\varepsilon^2$	$-\varepsilon^{2*}$	+1	$-\varepsilon^2$	$-\varepsilon^{2*}$			
E_5	+1	$-\varepsilon^*$	ε^{2*}	+i	$-\varepsilon^2$	ε	-1	ε^*	$-\varepsilon^{2*}$	-i	ε^2	$-\varepsilon$	R_x-iR_y, R_x+iR_y	(xz, yz)	
	+1	$-\varepsilon$	ε^2	-i	$-\varepsilon^{2*}$	ε^*	-1	ε	$-\varepsilon^2$	+i	ε^{2*}	$-\varepsilon^*$			

The C_{nh} groups

C_{2h}	E	$C_2(z)$	i	σ_h	(h = 4)		
A_g	+1	+1	+1	+1	R_z	x^2, y^2, z^2, xy	
B_g	+1	-1	+1	-1	R_x, R_y	xz, yz	
A_u	+1	+1	-1	-1	z		z^3, xyz, x^2z, y^2z
B_u	+1	-1	-1	+1	x, y		xz^2, yz^2, x^2y, xy^2, x^3, y^3

C_{3h}	E	$C_3(z)$	C_3^2	σ_h	S_3	S_3^5	(h = 6);	$\varepsilon = \exp(2\pi i/3)$	
A'	+1	+1	+1	+1	+1	+1	R_z	x^2+y^2, z^2	$y(3x^2-y^2)$, $x(x^2-3y^2)$
E'	+1	ε	ε^*	+1	ε	ε^*	x+iy, x-iy	(x^2-y^2, xy)	(xz^2, yz^2) $[x(x^2+y^2), y(x^2+y^2)]$
	+1	ε^*	ε	+1	ε^*	ε			
A"	+1	+1	+1	-1	-1	-1	z		z^3, $z(x^2+y^2)$
E"	+1	ε	ε^*	-1	$-\varepsilon$	$-\varepsilon^*$	R_x+iR_y, R_x-iR_y	(xz, yz)	$[xyz, z(x^2-y^2)]$
	+1	ε^*	ε	-1	$-\varepsilon^*$	$-\varepsilon$			

C_{4h}	E	$C_4(z)$	C_2	C_4^3	i	S_4^3	σ_h	S_4	(h = 8)		
A_g	+1	+1	+1	+1	+1	+1	+1	+1	R_z	x^2+y^2, z^2	
B_g	+1	-1	+1	-1	+1	-1	+1	-1		x^2-y^2, xy	
E_g	+1	+i	-1	-i	+1	+i	-1	-i	R_x+iR_y	(xz, yz)	
	+1	-i	-1	+i	+1	-i	-1	+i	R_x-iR_y		
A_u	+1	+1	+1	+1	-1	-1	-1	-1	z		z^3, $z(x^2+y^2)$
B_u	+1	-1	+1	-1	-1	+1	-1	+1			xyz, $z(x^2-y^2)$
E_u	+1	+i	-1	-i	-1	-i	+1	+i	x+iy		(xz^2, yz^2) (xy^2, x^2y) (x^3, y^3)
	+1	-i	-1	+i	-1	+i	+1	-i	x-iy		

C_{5h}	E	C_5	C_5^2	C_5^3	C_5^4	σ_h	S_5	S_5^7	S_5^3	S_5^9	(h = 10); $\varepsilon = \exp(2\pi i/5)$		
A'	+1	+1	+1	+1	+1	+1	+1	+1	+1	+1	R_z	x^2+y^2, z^2	
E_1'	+1	ε	ε^2	ε^{2*}	ε^*	+1	ε	ε^2	ε^{2*}	ε^*	x+iy		(xz^2, yz^2) $[x(x^2+y^2), y(x^2+y^2)]$
	+1	ε^*	ε^{2*}	ε^2	ε	+1	ε^*	ε^{2*}	ε^2	ε	x-iy		
E_2'	+1	ε^2	ε^*	ε	ε^{2*}	+1	ε^2	ε^*	ε	ε^{2*}		(x^2-y^2, xy)	$[y(3x^2-y^2), x(x^2-3y^2)]$
	+1	ε^{2*}	ε	ε^*	ε^2	+1	ε^{2*}	ε	ε^*	ε^2			
A"	+1	+1	+1	+1	+1	-1	-1	-1	-1	-1	z		z^3, $z(x^2+y^2)$
E_1''	+1	ε	ε^2	ε^{2*}	ε^*	-1	$-\varepsilon$	$-\varepsilon^2$	$-\varepsilon^{2*}$	$-\varepsilon^*$	R_x+iR_y	(xz, yz)	
	+1	ε^*	ε^{2*}	ε^2	ε	-1	$-\varepsilon^*$	$-\varepsilon^{2*}$	$-\varepsilon^2$	$-\varepsilon$	R_x-iR_y		
E_2''	+1	ε^2	ε^*	ε	ε^{2*}	-1	$-\varepsilon^2$	$-\varepsilon^*$	$-\varepsilon$	$-\varepsilon^{2*}$			$[xyz, z(x^2-y^2)]$
	+1	ε^{2*}	ε	ε^*	ε^2	-1	$-\varepsilon^{2*}$	$-\varepsilon$	$-\varepsilon^*$	$-\varepsilon^2$			

C_{6h}	E	$C_6(z)$	C_3	C_2	C_3^2	C_6^5	i	S_3^5	S_6^5	σ_h	S_6	S_3	(h = 12);	$\epsilon = \exp(2\pi i/6)$	
A_g	+1	+1	+1	+1	+1	+1	+1	+1	+1	+1	+1	+1	R_z	x^2+y^2, z^2	
B_g	+1	-1	+1	-1	+1	-1	+1	-1	+1	-1	+1	-1			
E_{1g}	+1	ϵ	$-\epsilon^*$	-1	$-\epsilon$	ϵ^*	+1	ϵ	$-\epsilon^*$	-1	$-\epsilon$	ϵ^*	R_x+iR_y	(xz, yz)	
	+1	ϵ^*	$-\epsilon$	-1	$-\epsilon^*$	ϵ	+1	ϵ^*	$-\epsilon$	-1	$-\epsilon^*$	ϵ	R_x-iR_y		
E_{2g}	+1	$-\epsilon^*$	$-\epsilon$	+1	$-\epsilon^*$	$-\epsilon$	+1	$-\epsilon^*$	$-\epsilon$	+1	$-\epsilon^*$	$-\epsilon$		(x^2-y^2, xy)	
	+1	$-\epsilon$	$-\epsilon^*$	+1	$-\epsilon$	$-\epsilon^*$	+1	$-\epsilon$	$-\epsilon^*$	+1	$-\epsilon$	$-\epsilon^*$			
A_u	+1	+1	+1	+1	+1	+1	-1	-1	-1	-1	-1	-1	z		$z^3, z(x^2+y^2)$
B_u	+1	-1	+1	-1	+1	-1	-1	+1	-1	+1	-1	+1			$y(3x^2-y^2), x(x^2-3y^2)$
E_{1u}	+1	ϵ	$-\epsilon^*$	-1	$-\epsilon$	ϵ^*	-1	$-\epsilon$	ϵ^*	+1	ϵ	$-\epsilon^*$	x+iy		$(xz^2, yz^2)[x(x^2+y^2), y(x^2+y^2)]$
	+1	ϵ^*	$-\epsilon$	-1	$-\epsilon^*$	ϵ	-1	$-\epsilon^*$	ϵ	+1	ϵ^*	$-\epsilon$	x-iy		
E_{2u}	+1	$-\epsilon^*$	$-\epsilon$	+1	$-\epsilon^*$	$-\epsilon$	-1	ϵ^*	ϵ	-1	ϵ^*	ϵ			$(xyz, z(x^2-y^2))$
	+1	$-\epsilon$	$-\epsilon^*$	+1	$-\epsilon$	$-\epsilon^*$	-1	ϵ	ϵ^*	-1	ϵ	ϵ^*			

The C_{nv} groups

C_{2v}	E	$C_2(z)$	$\sigma_v(xz)$	$\sigma'_v(yz)$	(h = 4)		
A_1	+1	+1	+1	+1	z	x^2, y^2, z^2	z^3, x^2z, y^2z
A_2	+1	+1	-1	-1	R_z	xy	xyz
B_1	+1	-1	+1	-1	x, R_y	xz	xz^2, x^3, xy^2
B_2	+1	-1	-1	+1	y, R_x	yz	yz^2, y^3, x^2y

C_{3v}	E	$2C_3(z)$	$3\sigma_v$	(h = 6)		
A_1	+1	+1	+1	z	x^2+y^2, z^2	$z^3, x(x^2-3y^2), z(x^2+y^2)$
A_2	+1	+1	-1	R_z		$y(3x^2-y^2)$
E	+2	-1	0	$(x, y) (R_x, R_y)$	$(x^2-y^2, xy) (xz, yz)$	$(xz^2, yz^2) [xyz, z(x^2-y^2)] [x(x^2+y^2), y(x^2+y^2)]$

C_{4v}	E	$2C_4(z)$	C_2	$2\sigma_v$	$2\sigma_d$	(h = 8) (x axis in σ_v plane)		
A_1	+1	+1	+1	+1	+1	z	x^2+y^2, z^2	z^3, $z(x^2+y^2)$
A_2	+1	+1	+1	-1	-1	R_z		
B_1	+1	-1	+1	+1	-1		x^2-y^2	$z(x^2-y^2)$
B_2	+1	-1	+1	-1	+1		xy	xyz
E	+2	0	-2	0	0	(x, y) (R_x, R_y)	(xz, yz)	(xz^2, yz^2) (xy^2, x^2y) (x^3, y^3)

C_{5v}	E	$2C_5(z)$	$2C_5^2$	$5\sigma_v$	(h = 10) τ - see §2.1		
A_1	+1	+1	+1	+1	z	x^2+y^2, z^2	z^3, $z(x^2+y^2)$
A_2	+1	+1	+1	-1	R_z		
E_1	+2	τ-1	$-\tau$	0	(x, y) (R_x, R_y)	(xz, yz)	(xz^2, yz^2) $[x(x^2+y^2), y(x^2+y^2)]$
E_2	+2	$-\tau$	τ-1	0		(x^2-y^2, xy)	$[xyz, z(x^2-y^2)]$ $[y(3x^2-y^2), x(x^2-3y^2)]$

C_{6v}	E	$2C_6(z)$	$2C_3(z)$	$C_2(z)$	$3\sigma_v$	$3\sigma_d$	(h = 12) (x axis in σ_v plane)		
A_1	+1	+1	+1	+1	+1	+1	z	x^2+y^2, z^2	z^3, $z(x^2+y^2)$
A_2	+1	+1	+1	+1	-1	-1	R_z		
B_1	+1	-1	+1	-1	+1	-1			$x(x^2-3y^2)$
B_2	+1	-1	+1	-1	-1	+1			$y(3x^2-y^2)$
E_1	+2	+1	-1	-2	0	0	(x, y) (R_x, R_y)	(xz, yz)	(xz^2, yz^2) $[x(x^2+y^2), y(x^2+y^2)]$
E_2	+2	-1	-1	+2	0	0		(x^2-y^2, xy)	$[xyz, z(x^2-y^2)]$

C_{7v}	E	$2C_7$	$2C_7^2$	$2C_7^3$	$7\sigma_v$	(h = 14)		
A_1	+1	+1	+1	+1	+1	z	z^2, x^2+y^2	z^3, $z(x^2+y^2)$
A_2	+1	+1	+1	+1	-1	R_z		
E_1	+2	$2\cos\frac{2}{7}\pi$	$2\cos\frac{4}{7}\pi$	$2\cos\frac{6}{7}\pi$	0	(x, y) (R_x, R_y)	(xz, yz)	(xz^2, yz^2) [$x(x^2+y^2)$, $y(x^2+y^2)$]
E_2	+2	$2\cos\frac{4}{7}\pi$	$2\cos\frac{6}{7}\pi$	$2\cos\frac{2}{7}\pi$	0		(x^2-y^2, xy)	[xyz, $z(x^2-y^2)$]
E_3	+2	$2\cos\frac{6}{7}\pi$	$2\cos\frac{2}{7}\pi$	$2\cos\frac{4}{7}\pi$	0			[$y(3x^2-y^2)$, $x(x^2-3y^2)$]

C_{8v}	E	$2C_8$	$2C_4$	$2C_8^3$	C_2	$4\sigma_v$	$4\sigma_d$	(h = 16) (x axis in σ_v plane)		
A_1	+1	+1	+1	+1	+1	+1	+1	z	z^2, x^2+y^2	z^3, $z(x^2+y^2)$
A_2	+1	+1	+1	+1	+1	-1	-1	R_z		
B_1	+1	-1	+1	-1	+1	+1	-1			
B_2	+1	-1	+1	-1	+1	-1	+1			
E_1	+2	$+\sqrt{2}$	0	$-\sqrt{2}$	-2	0	0	(x, y) (R_x, R_y)	(xz, yz)	(xz^2, yz^2) [$x(x^2+y^2)$, $y(x^2+y^2)$]
E_2	+2	0	-2	0	+2	0	0		(x^2-y^2, xy)	[xyz, $z(x^2-y^2)$]
E_3	+2	$-\sqrt{2}$	0	$+\sqrt{2}$	-2	0	0			[$y(3x^2-y^2)$, $x(x^2-3y^2)$]

The dihedral groups

D_2	E	$C_2(z)$	$C_2(y)$	$C_2(x)$	(h = 4)		
A	+1	+1	+1	+1		x^2, y^2, z^2	xyz
B_1	+1	+1	-1	-1	z, R_z	xy	z^3, y^2z, x^2z
B_2	+1	-1	+1	-1	y, R_y	xz	yz^2, x^2y, y^3
B_3	+1	-1	-1	+1	x, R_x	yz	xz^2, xy^2, x^3

D_3	E	$2C_3(z)$	$3C'_2$	(h = 6) (x axis coincident with C'_2 axis)		
A_1	+1	+1	+1		x^2+y^2, z^2	$x(x^2-3y^2)$
A_2	+1	+1	-1	z, R_z		z^3, $y(3x^2-y^2)$, $z(x^2+y^2)$
E	+2	-1	0	(x, y) (R_x, R_y)	(x^2-y^2, xy) (xz, yz)	(xz^2, yz^2) [xyz, $z(x^2-y^2)$] [$x(x^2+y^2)$, $y(x^2+y^2)$]

D_4	E	$2C_4(z)$	$C_2(\equiv C_4^2)$	$2C'_2$	$2C''_2$	(h = 8) (x axis coincident with C'_2 axis)		
A_1	+1	+1	+1	+1	+1		x^2+y^2, z^2	
A_2	+1	+1	+1	-1	-1	z, R_z		z^3, $z(x^2+y^2)$
B_1	+1	-1	+1	+1	-1		x^2-y^2	xyz
B_2	+1	-1	+1	-1	+1		xy	$z(x^2-y^2)$
E	+2	0	-2	0	0	(x, y) (R_x, R_y)	(xz, yz)	(xz^2, yz^2) (xy^2, x^2y) (x^3, y^3)

D_5	E	$2C_5(z)$	$2C_5^2$	$5C'_2$	(h = 10) τ - see §2.1 (x axis coincident with C'_2 axis)		
A_1	+1	+1	+1	+1		x^2+y^2, z^2	
A_2	+1	+1	+1	-1	z, R_z		z^3, $z(x^2+y^2)$
E_1	+2	τ-1	-τ	0	(x, y) (R_x, R_y)	(xz, yz)	(xz^2, yz^2) [$x(x^2+y^2)$, $y(x^2+y^2)$]
E_2	+2	-τ	τ-1	0		(x^2-y^2, xy)	[xyz, $z(x^2-y^2)$] [$y(3x^2-y^2)$, $x(x^2-3y^2)$]

D_6	E	$2C_6(z)$	$2C_3$	$C_2(z)$	$3C'_2$	$3C''_2$	(h = 12) (x axis coincident with C'_2 axis)		
A_1	+1	+1	+1	+1	+1	+1		x^2+y^2, z^2	
A_2	+1	+1	+1	+1	-1	-1	z, R_z		z^3, $z(x^2+y^2)$
B_1	+1	-1	+1	-1	+1	-1			$x(x^2-3y^2)$
B_2	+1	-1	+1	-1	-1	+1			$y(3x^2-y^2)$
E_1	+2	+1	-1	-2	0	0	(x, y) (R_x, R_y)	(xz, yz)	(xz^2, yz^2) [$x(x^2+y^2)$, $y(x^2+y^2)$]
E_2	+2	-1	-1	2	0	0		(x^2-y^2, xy)	[xyz, $z(x^2-y^2)$]

D_7	E	$2C_7$	$2C_7^2$	$2C_7^3$	$7C'_2$	(h = 14) (x axis coincident with C'_2 axis)		
A_1	+1	+1	+1	+1	+1		z^2, x^2+y^2	
A_2	+1	+1	+1	+1	-1	z, R_z		z^3, $z(x^2+y^2)$
E_1	+2	$2\cos\frac{2}{7}\pi$	$2\cos\frac{4}{7}\pi$	$2\cos\frac{6}{7}\pi$	0	(x, y); (R_x, R_y)	(xz, yz)	(xz^2, yz^2) [$x(x^2+y^2)$, $y(x^2+y^2)$]
E_2	+2	$2\cos\frac{4}{7}\pi$	$2\cos\frac{6}{7}\pi$	$2\cos\frac{2}{7}\pi$	0		(x^2-y^2, xy)	[xyz, $z(x^2-y^2)$]
E_3	+2	$2\cos\frac{6}{7}\pi$	$2\cos\frac{2}{7}\pi$	$2\cos\frac{4}{7}\pi$	0			[$y(3x^2-y^2)$, $x(x^2-3y^2)$]

D_8	E	$2C_8$	$2C_4$(C_8^2)	$2C_8^3$	C_2(C_8^4)	$4C'_2$	$4C''_2$	(h = 16) (x axis coincident with C'_2 axis)		
A_1	+1	+1	+1	+1	+1	+1	+1		z^2, x^2+y^2	
A_2	+1	+1	+1	+1	+1	-1	-1	z, R_z		z^3, $z(x^2+y^2)$
B_1	+1	-1	+1	-1	+1	+1	-1			
B_2	+1	-1	+1	-1	+1	-1	+1			
E_1	+2	$+\sqrt{2}$	0	$-\sqrt{2}$	-2	0	0	(x, y) (R_x, R_y)	(xz, yz)	(xz^2, yz^2) [$x(x^2+y^2)$, $y(x^2+y^2)$]
E_2	+2	0	-2	0	+2	0	0		(x^2-y^2, xy)	[xyz, $z(x^2-y^2)$]
E_3	+2	$-\sqrt{2}$	0	$+\sqrt{2}$	-2	0	0			[$y(3x^2-y^2)$, $x(x^2-3y^2)$]

D_{2h}	E	$C_2(z)$	$C_2(y)$	$C_2(x)$	i	σ(xy)	σ(xz)	σ(yz)	(h = 8)		
A_g	+1	+1	+1	+1	+1	+1	+1	+1		x^2, y^2, z^2	
B_{1g}	+1	+1	-1	-1	+1	+1	-1	-1	R_z	xy	
B_{2g}	+1	-1	+1	-1	+1	-1	+1	-1	R_y	xz	
B_{3g}	+1	-1	-1	+1	+1	-1	-1	+1	R_x	yz	
A_u	+1	+1	+1	+1	-1	-1	-1	-1			xyz
B_{1u}	+1	+1	-1	-1	-1	-1	+1	+1	z		z^3, y^2z, xz^2
B_{2u}	+1	-1	+1	-1	-1	+1	-1	+1	y		yz^2, x^2y, y^3
B_{3u}	+1	-1	-1	+1	-1	+1	+1	-1	x		xz^2, xy^2, x^3

D_{3h}	E	$2C_3(z)$	$3C'_2$	$\sigma_h(xy)$	$2S_3$	$3\sigma_v$	(h = 12) (x axis coincident with C'_2 axis)		
A'_1	+1	+1	+1	+1	+1	+1		x^2+y^2, z^2	$x(x^2-3y^2)$
A'_2	+1	+1	-1	+1	+1	-1	R_z		$y(3x^2-y^2)$
E'	+2	-1	0	+2	-1	0	(x, y)	(x^2-y^2, xy)	(xz^2, yz^2) $[x(x^2+y^2), y(x^2+y^2)]$
A''_1	+1	+1	+1	-1	-1	-1			
A''_2	+1	+1	-1	-1	-1	+1	z		z^3, $z(x^2+y^2)$
E"	+2	-1	0	-2	+1	0	(R_x, R_y)	(xz, yz)	$[xyz, z(x^2-y^2)]$

D_{4h}	E	$2C_4(z)$	C_2	$2C'_2$	$2C''_2$	i	$2S_4$	σ_h	$2\sigma_v$	$2\sigma_d$	(h = 16) (x axis coincident with C'_2 axis)		
A_{1g}	+1	+1	+1	+1	+1	+1	+1	+1	+1	+1		x^2+y^2, z^2	
A_{2g}	+1	+1	+1	-1	-1	+1	+1	+1	-1	-1	R_z		
B_{1g}	+1	-1	+1	+1	-1	+1	-1	+1	+1	-1		x^2-y^2	
B_{2g}	+1	-1	+1	-1	+1	+1	-1	+1	-1	+1		xy	
E_g	+2	0	-2	0	0	+2	0	-2	0	0	(R_x, R_y)	(xz, yz)	
A_{1u}	+1	+1	+1	+1	+1	-1	-1	-1	-1	-1			
A_{2u}	+1	+1	+1	-1	-1	-1	-1	-1	+1	+1	z		z^3, $z(x^2+y^2)$
B_{1u}	+1	-1	+1	+1	-1	-1	+1	-1	-1	+1			xyz
B_{2u}	+1	-1	+1	-1	+1	-1	+1	-1	+1	-1			$z(x^2-y^2)$
E_u	+2	0	-2	0	0	-2	0	+2	0	0	(x, y)		(xz^2, yz^2) (xy^2, x^2y) (x^3, y^3)

D_{5h}	E	$2C_5$	$2C_5^2$	$5C'_2$	σ_h	$2S_5$	$2S_5^3$	$5\sigma_v$	(h = 20) τ - see §2.1 (x axis coincident with C'_2 axis)		
A'_1	+1	+1	+1	+1	+1	+1	+1	+1		x^2+y^2, z^2	
A'_2	+1	+1	+1	-1	+1	+1	+1	-1	R_z		
E'_1	+2	τ-1	-τ	0	+2	τ-1	-τ	0	(x, y)		(xz^2, yz^2) $[x(x^2+y^2), y(x^2+y^2)]$
E'_2	+2	-τ	τ-1	0	+2	-τ	τ-1	0		(x^2-y^2, xy)	$[y(3x^2-y^2), x(x^2-3y^2)]$
A''_1	+1	+1	+1	+1	-1	-1	-1	-1			
A''_2	+1	+1	+1	-1	-1	-1	-1	+1	z		z^3, $z(x^2+y^2)$
E''_1	+2	τ-1	-τ	0	-2	1-τ	+τ	0	(R_x, R_y)	(xz, yz)	
E''_2	+2	-τ	τ-1	0	-2	+τ	1-τ	0			$[xyz, z(x^2-y^2)]$

D_{6h}	E	$2C_6(z)$	$2C_3$	C_2	$3C'_2$	$3C''_2$	i	$2S_3$	$2S_6$	$\sigma_h(xy)$	$3\sigma_d$	$3\sigma_v$	(h = 24) (x axis coincident with C'_2 axis)		
A_{1g}	+1	+1	+1	+1	+1	+1	+1	+1	+1	+1	+1	+1		x^2+y^2, z^2	
A_{2g}	+1	+1	+1	+1	-1	-1	+1	+1	+1	+1	-1	-1	R_z		
B_{1g}	+1	-1	+1	-1	+1	-1	+1	-1	+1	-1	+1	-1			
B_{2g}	+1	-1	+1	-1	-1	+1	+1	-1	+1	-1	-1	+1			
E_{1g}	+2	+1	-1	-2	0	0	+2	+1	-1	-2	0	0	(R_x, R_y)	(xz, yz)	
E_{2g}	+2	-1	-1	+2	0	0	+2	-1	-1	+2	0	0		(x^2-y^2, xy)	
A_{1u}	+1	+1	+1	+1	+1	+1	-1	-1	-1	-1	-1	-1			
A_{2u}	+1	+1	+1	+1	-1	-1	-1	-1	-1	-1	+1	+1	z		z^3, $z(x^2+y^2)$
B_{1u}	+1	-1	+1	-1	+1	-1	-1	+1	-1	+1	-1	+1			$x(x^2-3y^2)$
B_{2u}	+1	-1	+1	-1	-1	+1	-1	+1	-1	+1	+1	-1			$y(3x^2-y^2)$
E_{1u}	+2	+1	-1	-2	0	0	-2	-1	+1	+2	0	0	(x, y)		(xz^2, yz^2) $[x(x^2+y^2), y(x^2+y^2)]$
E_{2u}	+2	-1	-1	+2	0	0	-2	+1	+1	-2	0	0			$[xyz, z(x^2-y^2)]$

D_{7h}	E	$2C_7$	$2C_7^2$	$2C_7^3$	$7C'_2$	σ_h	$2S_7$	$2S_7^5$	$2S_7^3$	$7\sigma_v$	(h = 28)	(x axis coincident with C'_2 axis)	
A'_1	+1	+1	+1	+1	+1	+1	+1	+1	+1	+1		z^2, x^2+y^2	
A'_2	+1	+1	+1	+1	-1	+1	+1	+1	+1	-1	R_z		
E'_1	+2	$2\cos\frac{2}{7}\pi$	$2\cos\frac{4}{7}\pi$	$2\cos\frac{6}{7}\pi$	0	+2	$2\cos\frac{2}{7}\pi$	$2\cos\frac{4}{7}\pi$	$2\cos\frac{6}{7}\pi$	0	(x, y)		(xz^2, yz^2) $[x(x^2+y^2), y(x^2+y^2)]$
E'_2	+2	$2\cos\frac{4}{7}\pi$	$2\cos\frac{6}{7}\pi$	$2\cos\frac{2}{7}\pi$	0	+2	$2\cos\frac{4}{7}\pi$	$2\cos\frac{6}{7}\pi$	$2\cos\frac{2}{7}\pi$	0		(x^2-y^2, xy)	
E'_3	+2	$2\cos\frac{6}{7}\pi$	$2\cos\frac{2}{7}\pi$	$2\cos\frac{4}{7}\pi$	0	+2	$2\cos\frac{6}{7}\pi$	$2\cos\frac{2}{7}\pi$	$2\cos\frac{4}{7}\pi$	0			$[y(3x^2-y^2), x(x^2-3y^2)]$
A''_1	+1	+1	+1	+1	+1	-1	-1	-1	-1	-1			
A''_2	+1	+1	+1	+1	-1	-1	-1	-1	-1	+1	z		z^3, $z(x^2+y^2)$
E''_1	+2	$2\cos\frac{2}{7}\pi$	$2\cos\frac{4}{7}\pi$	$2\cos\frac{6}{7}\pi$	0	-2	$-2\cos\frac{2}{7}\pi$	$-2\cos\frac{4}{7}\pi$	$-2\cos\frac{6}{7}\pi$	0	(R_x, R_y)	(xz, yz)	
E''_2	+2	$2\cos\frac{4}{7}\pi$	$2\cos\frac{6}{7}\pi$	$2\cos\frac{2}{7}\pi$	0	-2	$-2\cos\frac{4}{7}\pi$	$-2\cos\frac{6}{7}\pi$	$-2\cos\frac{2}{7}\pi$	0			$[xyz, z(x^2-y^2)]$
E''_3	+2	$2\cos\frac{6}{7}\pi$	$2\cos\frac{2}{7}\pi$	$2\cos\frac{4}{7}\pi$	0	-2	$-2\cos\frac{6}{7}\pi$	$-2\cos\frac{2}{7}\pi$	$-2\cos\frac{4}{7}\pi$	0			

D_{8h}	E	$2C_8$	$2C_8^3$	$2C_4$	$C_2(z)$	$4C'_2$	$4C''_2$	i	$2S_8^3$	$2S_8$	$2S_4$	σ_h	$4\sigma_v$	$4\sigma_d$	(h = 32)	(x axis coincident with C'_2 axis)	
A_{1g}	+1	+1	+1	+1	+1	+1	+1	+1	+1	+1	+1	+1	+1	+1		z^2, x^2+y^2	
A_{2g}	+1	+1	+1	+1	+1	-1	-1	+1	+1	+1	+1	+1	-1	-1	R_z		
B_{1g}	+1	-1	-1	+1	+1	+1	-1	+1	-1	-1	+1	+1	+1	-1			
B_{2g}	+1	-1	-1	+1	+1	-1	+1	+1	-1	-1	+1	+1	-1	+1			
E_{1g}	+2	$+\sqrt{2}$	$-\sqrt{2}$	0	-2	0	0	+2	$+\sqrt{2}$	$-\sqrt{2}$	0	-2	0	0	(R_x, R_y)	(xz, yz)	
E_{2g}	+2	0	0	-2	+2	0	0	+2	0	0	-2	+2	0	0		(x^2-y^2, xy)	
E_{3g}	+2	$-\sqrt{2}$	$+\sqrt{2}$	0	-2	0	0	+2	$-\sqrt{2}$	$+\sqrt{2}$	0	-2	0	0			
A_{1u}	+1	+1	+1	+1	+1	+1	+1	-1	-1	-1	-1	-1	-1	-1			
A_{2u}	+1	+1	+1	+1	+1	-1	-1	-1	-1	-1	-1	-1	+1	+1	z		z^3, $z(x^2+y^2)$
B_{1u}	+1	-1	-1	+1	+1	+1	-1	-1	+1	+1	-1	-1	-1	+1			
B_{2u}	+1	-1	-1	+1	+1	-1	+1	-1	+1	+1	-1	-1	+1	-1			
E_{1u}	+2	$+\sqrt{2}$	$-\sqrt{2}$	0	-2	0	0	-2	$-\sqrt{2}$	$+\sqrt{2}$	0	+2	0	0	(x, y)		(xz^2, yz^2) $[x(x^2+y^2), y(x^2+y^2)]$
E_{2u}	+2	0	0	-2	+2	0	0	-2	0	0	+2	-2	0	0			$[xyz, z(x^2-y^2)]$
E_{3u}	+2	$-\sqrt{2}$	$+\sqrt{2}$	0	-2	0	0	-2					0	0			$[y(3x^2-y^2), x(x^2-3y^2)]$

D_{2d}	E	$2S_4$	$C_2(z)$	$2C'_2$	$2\sigma_d$	(h = 8) (x axis coincident with C'_2 axis)		
A_1	+1	+1	+1	+1	+1		x^2+y^2, z^2	xyz
A_2	+1	+1	+1	-1	-1	R_z		$z(x^2-y^2)$
B_1	+1	-1	+1	+1	-1		x^2-y^2	
B_2	+1	-1	+1	-1	+1	z	xy	$z^3, z(x^2+y^2)$
E	+2	0	-2	0	0	(x, y) (R_x, R_y)	(xz, yz)	(xz^2, yz^2) (xy^2, x^2y) (x^3, y^3)

D_{3d}	E	$2C_3$	$3C'_2$	i	$2S_6$	$3\sigma_d$	(h = 12) (x axis coincident with C'_2 axis)		
A_{1g}	+1	+1	+1	+1	+1	+1		x^2+y^2, z^2	
A_{2g}	+1	+1	-1	+1	+1	-1	R_z		
E_g	+2	-1	0	+2	-1	0	(R_x, R_y)	(x^2-y^2, xy) (xz, yz)	
A_{1u}	+1	+1	+1	-1	-1	-1			$x(x^2-3y^2)$
A_{2u}	+1	+1	-1	-1	-1	+1	z		$y(3x^2-y^2), z^3, z(x^2+y^2)$
E_u	+2	-1	0	-2	+1	0	(x, y)		(xz^2, yz^2) $[xyz, z(x^2-y^2)]$ $[x(x^2+y^2), y(x^2+y^2)]$

D_{4d}	E	$2S_8$	$2C_4$	$2S_8^3$	C_2	$4C'_2$	$4\sigma_d$	(h = 16) (x axis coincident with C'_2 axis)		
A_1	+1	+1	+1	+1	+1	+1	+1		x^2+y^2, z^2	
A_2	+1	+1	+1	+1	+1	-1	-1	R_z		
B_1	+1	-1	+1	-1	+1	+1	-1			
B_2	+1	-1	+1	-1	+1	-1	+1	z		$z^3, z(x^2+y^2)$
E_1	+2	$+\sqrt{2}$	0	$-\sqrt{2}$	-2	0	0	(x, y)		(xz^2, yz^2) $[x(x^2+y^2), y(x^2+y^2)]$
E_2	+2	0	-2	0	+2	0	0		(x^2-y^2, xy)	$[xyz, z(x^2-y^2)]$
E_3	+2	$-\sqrt{2}$	0	$+\sqrt{2}$	-2	0	0	(R_x, R_y)	(xz, yz)	$[y(3x^2-y^2), x(x^2-3y^2)]$

D_{5d}	E	$2C_5$	$2C_5^2$	$5C'_2$	i	$2S_{10}^3$	$2S_{10}$	$5\sigma_d$	(h = 20) τ - see §2.1 (x axis coincident with C'_2 axis)		
A_{1g}	+1	+1	+1	+1	+1	+1	+1	+1		x^2+y^2, z^2	
A_{2g}	+1	+1	+1	-1	+1	+1	+1	-1	R_z		
E_{1g}	+2	τ-1	-τ	0	+2	τ-1	-τ	0	(R_x, R_y)	(xz, yz)	
E_{2g}	+2	-τ	τ-1	0	+2	-τ	τ-1	0		(x^2-y^2, xy)	
A_{1u}	+1	+1	+1	+1	-1	-1	-1	-1			
A_{2u}	+1	+1	+1	-1	-1	-1	-1	+1	z		z^3, $z(x^2+y^2)$
E_{1u}	+2	τ-1	-τ	0	-2	1-τ	+τ	0	(x, y)		(xz^2, yz^2) $[x(x^2+y^2), y(x^2+y^2)]$
E_{2u}	+2	-τ	τ-1	0	-2	+τ	1-τ	0			$[xyz, z(x^2-y^2)]$ $[y(3x^2-y^2), x(x^2-3y^2)]$

D_{6d}	E	$2S_{12}$	$2C_6$	$2S_4$	$2C_3$	$2S_{12}^5$	C_2	$6C'_2$	$6\sigma_d$	(h = 24) (x axis coincident with C'_2 axis)		
A_1	+1	+1	+1	+1	+1	+1	+1	+1	+1		x^2+y^2, z^2	
A_2	+1	+1	+1	+1	+1	+1	+1	-1	-1	R_z		
B_1	+1	-1	+1	-1	+1	-1	+1	+1	-1			
B_2	+1	-1	+1	-1	+1	-1	+1	-1	+1	z		z^3, $z(x^2+y^2)$
E_1	+2	+$\sqrt{3}$	+1	0	-1	-$\sqrt{3}$	-2	0	0	(x, y)		(xz^2, yz^2) $[x(x^2+y^2), y(x^2+y^2)]$
E_2	+2	+1	-1	-2	-1	+1	+2	0	0		(x^2-y^2, xy)	
E_3	+2	0	-2	0	+2	0	-2	0	0			$[y(3x^2-y^2), x(x^2-3y^2)]$
E_4	+2	-1	-1	+2	-1	-1	+2	0	0			$[xyz, z(x^2-y^2)]$
E_5	+2	-$\sqrt{3}$	+1	0	-1	+$\sqrt{3}$	-2	0	0	(R_x, R_y)	(xz, yz)	

D_{7d}	E	$2C_7$	$2C_7^2$	$2C_7^3$	$7C'_2$	i	$2S_{14}^5$	$2S_{14}^3$	$2S_{14}$	$7\sigma_d$	(h = 28)	(x axis coincident with C'_2 axis)	
A_{1g}	+1	+1	+1	+1	+1	+1	+1	+1	+1	+1		z^2, x^2+y^2	
A_{2g}	+1	+1	+1	+1	-1	+1	+1	+1	+1	-1	R_z		
E_{1g}	+2	$2\cos\frac{2}{7}\pi$	$2\cos\frac{4}{7}\pi$	$2\cos\frac{6}{7}\pi$	0	+2	$2\cos\frac{2}{7}\pi$	$2\cos\frac{4}{7}\pi$	$2\cos\frac{6}{7}\pi$	0	(R_x, R_y)	(xz, yz)	
E_{2g}	+2	$2\cos\frac{4}{7}\pi$	$2\cos\frac{6}{7}\pi$	$2\cos\frac{2}{7}\pi$	0	+2	$2\cos\frac{4}{7}\pi$	$2\cos\frac{6}{7}\pi$	$2\cos\frac{2}{7}\pi$	0		(x^2-y^2, xy)	
E_{3g}	+2	$2\cos\frac{6}{7}\pi$	$2\cos\frac{2}{7}\pi$	$2\cos\frac{4}{7}\pi$	0	+2	$2\cos\frac{6}{7}\pi$	$2\cos\frac{2}{7}\pi$	$2\cos\frac{4}{7}\pi$	0			
A_{1u}	+1	+1	+1	+1	+1	-1	-1	-1	-1	-1			
A_{2u}	+1	+1	+1	+1	-1	-1	-1	-1	-1	+1	z		$z^3, z(x^2+y^2)$
E_{1u}	+2	$2\cos\frac{2}{7}\pi$	$2\cos\frac{4}{7}\pi$	$2\cos\frac{6}{7}\pi$	0	-2	$-2\cos\frac{2}{7}\pi$	$-2\cos\frac{4}{7}\pi$	$-2\cos\frac{6}{7}\pi$	0	(x, y)		$(xz^2, yz^2)[x(x^2+y^2), y(x^2+y^2)]$
E_{2u}	+2	$2\cos\frac{4}{7}\pi$	$2\cos\frac{6}{7}\pi$	$2\cos\frac{2}{7}\pi$	0	-2	$-2\cos\frac{4}{7}\pi$	$-2\cos\frac{6}{7}\pi$	$-2\cos\frac{2}{7}\pi$	0			$[xyz, z(x^2-y^2)]$
E_{3u}	+2	$2\cos\frac{6}{7}\pi$	$2\cos\frac{2}{7}\pi$	$2\cos\frac{4}{7}\pi$	0	-2	$-2\cos\frac{6}{7}\pi$	$-2\cos\frac{2}{7}\pi$	$-2\cos\frac{4}{7}\pi$	0			$[y(3x^2-y^2), x(x^2-3y^2)]$

D_{8d}	E	$2S_{16}$	$2C_8$	$2S_{16}^3$	$2C_4$	$2S_{16}^5$	$2C_8^3$	$2S_{16}^7$	$C_2(z)$	$8C'_2$	$8\sigma_d$	(h = 32)	(x axis coincident with C'_2 axis)	
A_1	+1	+1	+1	+1	+1	+1	+1	+1	+1	+1	+1		z^2, x^2+y^2	
A_2	+1	+1	+1	+1	+1	+1	+1	+1	+1	-1	-1	R_z		
B_1	+1	-1	+1	-1	+1	-1	+1	-1	+1	+1	-1			
B_2	+1	-1	+1	-1	+1	-1	+1	-1	+1	-1	+1	z		$z^3, z(x^2+y^2)$
E_1	+2	$2\cos\frac{1}{8}\pi$	$+\sqrt{2}$	$2\cos\frac{3}{8}\pi$	0	$-2\cos\frac{3}{8}\pi$	$-\sqrt{2}$	$-2\cos\frac{1}{8}\pi$	-2	0	0	(x, y)		$(xz^2, yz^2)[x(x^2+y^2), y(x^2+y^2)]$
E_2	+2	$+\sqrt{2}$	0	$-\sqrt{2}$	-2	$-\sqrt{2}$	0	$+\sqrt{2}$	+2	0	0		(x^2-y^2, xy)	
E_3	+2	$2\cos\frac{3}{8}\pi$	$-\sqrt{2}$	$-2\cos\frac{1}{8}\pi$	0	$2\cos\frac{1}{8}\pi$	$+\sqrt{2}$	$-2\cos\frac{3}{8}\pi$	-2	0	0			$[x(x^2-3y^2), y(3x^2-y^2)]$
E_4	+2	0	-2	0	+2	0	-2	0	+2	0	0			
E_5	+2	$-2\cos\frac{3}{8}\pi$	$-\sqrt{2}$	$2\cos\frac{1}{8}\pi$	0	$-2\cos\frac{1}{8}\pi$	$+\sqrt{2}$	$2\cos\frac{3}{8}\pi$	-2	0	0			
E_6	+2	$-\sqrt{2}$	0	$+\sqrt{2}$	-2	$+\sqrt{2}$	0	$-\sqrt{2}$	+2	0	0			$[xyz, z(x^2-y^2)]$
E_7	+2	$-2\cos\frac{1}{8}\pi$	$+\sqrt{2}$	$-2\cos\frac{3}{8}\pi$	0	$2\cos\frac{3}{8}\pi$	$-\sqrt{2}$	$2\cos\frac{1}{8}\pi$	-2	0	0	(R_x, R_y)	(xz, yz)	

The linear groups

$C_{\infty v}$	E	$2C_\infty^\phi$	...	$\infty\sigma_v$	($h = \infty$)		
$A_1 \equiv \Sigma^+$	+1	+1	...	+1	z	x^2+y^2, z^2	z^3, $z(x^2+y^2)$
$A_2 \equiv \Sigma^-$	+1	+1	...	-1	R_z		
$E_1 \equiv \Pi$	+2	$2\cos\phi$	...	0	(x, y) (R_x, R_y)	(xz, yz)	(xz^2, yz^2) $[x(x^2+y^2), y(x^2+y^2)]$
$E_2 \equiv \Delta$	+2	$2\cos 2\phi$	...	0		(x^2-y^2, xy)	$[xyz, z(x^2-y^2)]$
$E_3 \equiv \Phi$	+2	$2\cos 3\phi$	...	0			$[y(3x^2-y^2), x(x^2-3y^2)]$
...	...	...	...	...			
E_n	+2	$2\cos n\phi$	...	0			

$D_{\infty h}$	E	$2C_\infty^\phi$	...	$\infty\sigma_v$	i	$2S_\infty^\phi$	...	$\infty C'_2$	($h = \infty$)		
$A_{1g} \equiv \Sigma_g^+$	+1	+1	...	+1	+1	+1	...	+1		x^2+y^2, z^2	
$A_{2g} \equiv \Sigma_g^-$	+1	+1	...	-1	+1	+1	...	-1	R_z		
$E_{1g} \equiv \Pi_g$	+2	$2\cos\phi$	...	0	+2	$-2\cos\phi$	...	0	(R_x, R_y)	(xz, yz)	
$E_{2g} \equiv \Delta_g$	+2	$2\cos 2\phi$	...	0	+2	$2\cos 2\phi$	...	0		(x^2-y^2, xy)	
$E_{3g} \equiv \Phi_g$	+2	$2\cos 3\phi$	...	0	+2	$-2\cos 3\phi$	...	0			
E_{ng}	+2	$2\cos n\phi$	...	0	+2	$(-1)^n 2\cos n\phi$	...	0			
...	...	...	...	...	...	...	...	...			
$A_{1u} \equiv \Sigma_u^+$	+1	+1	...	+1	-1	-1	...	-1	z		z^3, $z(x^2+y^2)$
$A_{2u} \equiv \Sigma_u^-$	+1	+1	...	-1	-1	-1	...	+1			
$E_{1u} \equiv \Pi_u$	+2	$2\cos\phi$	...	0	-2	$2\cos\phi$	...	0	(x, y)		(xz^2, yz^2) $[x(x^2+y^2), y(x^2+y^2)]$
$E_{2u} \equiv \Delta_u$	+2	$2\cos 2\phi$	...	0	-2	$-2\cos 2\phi$	...	0			$[xyz, z(x^2-y^2)]$
$E_{3u} \equiv \Phi_u$	+2	$2\cos 3\phi$	...	0	-2	$2\cos 3\phi$	...	0			$[y(3x^2-y^2), x(x^2-3y^2)]$
E_{nu}	+2	$2\cos n\phi$	...	0	-2	$(-1)^{n+1} 2\cos n\phi$	...	0			
...	...	...	...	...	...	...	...	...			

6.3 The cubic groups

T	E	$4C_3$	$4C_3^2$	$3C_2$	$(h = 12)$; $\varepsilon = \exp(2\pi i/3)$		
A	+1	+1	+1	+1		$x^2+y^2+z^2$	xyz
E	{ +1	$+\varepsilon$	$+\varepsilon^*$	+1 }		$(x^2-y^2, 2z^2-x^2-y^2)$	
	{ +1	$+\varepsilon^*$	$+\varepsilon$	+1 }			
F	+3	0	0	-1	(x, y, z), (R_x, R_y, R_z)	(xy, xz, yz)	(x^3, y^3, z^3) (xy^2, x^2z, yz^2) (xz^2, x^2y, y^2z)

T_h	E	$4C_3$	$4C_3^2$	$3C_2$	i	$4S_6^5$	$4S_6$	$3\sigma_h$	$(h = 24)$; $\varepsilon = \exp(2\pi i/3)$		
A_g	+1	+1	+1	+1	+1	+1	+1	+1		$x^2+y^2+z^2$	
E_g	{ +1	$+\varepsilon$	$+\varepsilon^*$	+1	+1	$+\varepsilon$	$+\varepsilon^*$	+1 }		$(x^2-y^2, 2z^2-x^2-y^2)$	
	{ +1	$+\varepsilon^*$	$+\varepsilon$	+1	+1	$+\varepsilon^*$	$+\varepsilon$	+1 }			
F_g	+3	0	0	-1	+3	0	0	-1	(R_x, R_y, R_z)	(xy, xz, yz)	
A_u	+1	+1	+1	+1	-1	-1	-1	-1			xyz
E_u	{ +1	$+\varepsilon$	$+\varepsilon^*$	+1	-1	$-\varepsilon$	$-\varepsilon^*$	-1 }			
	{ +1	$+\varepsilon^*$	$+\varepsilon$	+1	-1	$-\varepsilon^*$	$-\varepsilon$	-1 }			
F_u	+3	0	0	-1	-3	0	0	+1	(x, y, z)		(x^3, y^3, z^3) (xy^2, x^2z, yz^2) (xz^2, x^2y, y^2z)

T_d	E	$8C_3$	$3C_2$	$6S_4$	$6\sigma_d$	$(h = 24)$		
A_1	+1	+1	+1	+1	+1		$x^2+y^2+z^2$	xyz
A_2	+1	+1	+1	-1	-1			
E	+2	-1	+2	0	0		$(2z^2-x^2-y^2, x^2-y^2)$	
F_1	+3	0	-1	+1	-1	(R_x, R_y, R_z)		$[x(z^2-y^2), y(z^2-x^2), z(x^2-y^2)]$
F_2	+3	0	-1	-1	+1	(x, y, z)	(xy, xz, yz)	(x^3, y^3, z^3) $[x(z^2+y^2), y(z^2+x^2), z(x^2+y^2)]$

O	E	$8C_3$	$6C'_2$	$6C_4$	$3C_2(\equiv C_4^2)$	(h = 24)		
A_1	+1	+1	+1	+1	+1		$x^2+y^2+z^2$	
A_2	+1	+1	-1	-1	+1			xyz
E	+2	-1	0	0	+2		$(x^2-y^2,\ 2z^2-x^2-y^2)$	
F_1	+3	0	-1	+1	-1	(x, y, z) (R_x, R_y, R_z)		(x^3, y^3, z^3) $[x(z^2+y^2), y(z^2+x^2), z(x^2+y^2)]$
F_2	+3	0	+1	-1	-1		$(xy,\ xz,\ yz)$	$[x(z^2-y^2),\ y(z^2-x^2),\ z(x^2-y^2)]$

O_h	E	$8C_3$	$6C_2$	$6C_4$	$3C_2(\equiv C_4^2)$	i	$6S_4$	$8S_6$	$3\sigma_h$	$6\sigma_d$	(h = 48)		
A_{1g}	+1	+1	+1	+1	+1	+1	+1	+1	+1	+1		$x^2+y^2+z^2$	
A_{2g}	+1	+1	-1	-1	+1	+1	-1	+1	+1	-1			
E_g	+2	-1	0	0	+2	+2	0	-1	+2	0		$(2z^2-x^2-y^2, x^2-y^2)$	
F_{1g}	+3	0	-1	+1	-1	+3	+1	0	-1	-1	(R_x, R_y, R_z)		
F_{2g}	+3	0	+1	-1	-1	+3	-1	0	-1	+1		$(xz,\ yz,\ xy)$	
A_{1u}	+1	+1	+1	+1	+1	-1	-1	-1	-1	-1			
A_{2u}	+1	+1	-1	-1	+1	-1	+1	-1	-1	+1			xyz
E_u	+2	-1	0	0	+2	-2	0	+1	-2	0			
F_{1u}	+3	0	-1	+1	-1	-3	-1	0	+1	+1	$(x,\ y,\ z)$		(x^3, y^3, z^3) $[x(z^2+y^2), y(z^2+x^2), z(x^2+y^2)]$
F_{2u}	+3	0	+1	-1	-1	-3	+1	0	+1	-1			$[x(z^2-y^2),\ y(z^2-x^2),\ z(x^2-y^2)]$

6.4 The icosahedral groups

I	E	$12C_5$	$12C_5^2$	$20C_3$	$15C_2$	(h = 60) τ - see §2.1		
A	+1	+1	+1	+1	+1		$x^2+y^2+z^2$	
F_1	+3	$+\tau$	$1-\tau$	0	-1	(x, y, z) (R_x, R_y, R_z)		$[x(z^2+y^2), y(z^2+x^2), z(x^2+y^2)]$
F_2	+3	$1-\tau$	$+\tau$	0	-1			(x^3, y^3, z^3)
G	+4	-1	-1	+1	0			$[x(z^2-y^2), y(z^2-x^2), z(x^2-y^2), xyz]$
H	+5	0	0	-1	+1		$[2z^2-x^2-y^2, x^2-y^2, xy, xz, yz]$	

I_h	E	$12C_5$	$12C_5^2$	$20C_3$	$15C_2$	i	$12S_{10}$	$12S_{10}^3$	$20S_6$	15σ	(h = 120) τ - see §2.1		
A_g	+1	+1	+1	+1	+1	+1	+1	+1	+1	+1		$x^2+y^2+z^2$	
F_{1g}	+3	$+\tau$	$1-\tau$	0	-1	+3	$1-\tau$	$+\tau$	0	-1	(R_x, R_y, R_z)		
F_{2g}	+3	$1-\tau$	$+\tau$	0	-1	+3	$+\tau$	$1-\tau$	0	-1			
G_g	+4	-1	-1	+1	0	+4	-1	-1	+1	0			
H_g	+5	0	0	-1	+1	+5	0	0	-1	+1		$(2z^2-x^2-y^2, x^2-y^2, xy, xz, yz)$	
A_u	+1	+1	+1	+1	+1	-1	-1	-1	-1	-1			
F_{1u}	+3	$+\tau$	$1-\tau$	0	-1	-3	$\tau-1$	$-\tau$	0	+1	(x, y, z)		$[x(z^2+y^2), y(z^2+x^2), z(x^2+y^2)]$
F_{2u}	+3	$1-\tau$	$+\tau$	0	-1	-3	$-\tau$	$\tau-1$	0	+1			(x^3, y^3, z^3)
G_u	+4	-1	-1	+1	0	-4	+1	+1	-1	0			$[x(z^2-y^2), y(z^2-x^2), z(x^2-y^2), xyz]$
H_u	+5	0	0	-1	+1	-5	0	0	+1	-1			

6.5 The three-dimensional rotation groups R(3) and O(3)

The group **R(3)** comprises the infinite number of possible proper rotations about a point in three-dimensional space; i.e. it is the pure rotation subgroup of a sphere. The character, $\chi_j(\theta)$, of the irreducible representation, $D^{(j)}$, under a pure rotation through an angle θ is given by:

$$\chi_j(\theta) = \sum_{m=-j}^{m=+j} \exp(im\theta) = \frac{\sin \frac{1}{2}(2j+1)\theta}{\sin \frac{1}{2}\theta} . \quad [\text{When } \theta = 0, \ \chi_j(\theta) = (2j+1).]$$

When j is half-integral, the corresponding double group, **R(3)'**, is used. §6.6 lists the characters of some of the irreducible representations of **R(3)** and **R(3)'**. The full orthogonal group **O(3)** is the direct product group $\mathbf{C_i} \times \mathbf{R(3)}$, and has the full symmetry of the sphere.

6.6 Some double groups

In the following character tables for double groups, the identity operation, E, denotes rotation through an angle of 4π radians and the operation, R, rotation through 2π radians. A symmetry operation consisting of a rotation through $(2\pi + \theta)$ radians, where $\theta = \frac{2\pi k}{n}$, is written $C_n^k R$. For the characters of the irreducible representations common to a point group **G** and its corresponding double group **G'** (the 'single-valued' representations of **G**):

$$\chi_i(C_n^k R) = \chi_i(C_n^k) .$$

For the characters of the new irreducible representations of **G'** (the 'double-valued' representations of **G**):

$$\chi_j(C_n^k R) = -\chi_j(C_n^k) .$$

If $C_n^k = C_2^1$, and **G** also contains another C_2 axis perpendicular to the first:

$$\chi_j(C_2 R) = \chi_j(C_2) = 0 .$$

$\mathbf{T_d}$	E	R	$4C_3$	$4C_3^2$	$3C_2$	$3S_4$	$3S_4^3$	$6\sigma_d$
			$4C_3R$	$4C_3^2R$	$3C_2R$	$3S_4R$	$3S_4^3R$	$6\sigma_dR$
O	E	R	$4C_3$	$4C_3^2$	$3C_2$	$3C_4$	$3C_4^3$	$6C_2'$
(h=48)			$4C_3^2R$	$4C_3R$	$3C_2R$	$3C_4^3R$	$3C_4R$	$6C_2'R$
Γ_1 A_1	+1	+1	+1	+1	+1	+1	+1	+1
Γ_2 A_2	+1	+1	+1	+1	+1	-1	-1	-1
Γ_3 E_1	+2	+2	-1	-1	+2	0	0	0
Γ_4 T_1	+3	+3	0	0	-1	+1	+1	-1
Γ_5 T_2	+3	+3	0	0	-1	-1	-1	+1
Γ_6 $E_{\frac{1}{2}}$	+2	-2	+1	-1	0	$\sqrt{2}$	$-\sqrt{2}$	0
Γ_7 $E_{\frac{5}{2}}$	+2	-2	+1	-1	0	$-\sqrt{2}$	$\sqrt{2}$	0
Γ_8 $U_{\frac{3}{2}}$	+4	-4	-1	+1	0	0	0	0

T (h = 24)		E	R	$4C_3$	$4C_3R$	$4C_3^2$	$4C_3^2R$	$3C_2$ $3C_2R$
Γ_1	A	+1	+1	+1	+1	+1	+1	+1
Γ_2	E_1	+1	+1	$+\varepsilon$	$+\varepsilon$	$+\varepsilon^*$	$+\varepsilon^*$	+1
Γ_3		+1	+1	$+\varepsilon^*$	$+\varepsilon^*$	$+\varepsilon$	$+\varepsilon$	+1
Γ_4	T	+3	+3	0	0	0	0	-1
Γ_5	$E_{\frac{1}{2}}$	+2	-2	+1	-1	-1	+1	0
Γ_6	$U_{\frac{3}{2}}$	+2	-2	$+\varepsilon$	$-\varepsilon$	$-\varepsilon^*$	$+\varepsilon^*$	0
Γ_7		+2	-2	$+\varepsilon^*$	$-\varepsilon^*$	$-\varepsilon$	$+\varepsilon$	0

$\varepsilon = \exp(2\pi i/3)$

D_{3h}		E	R	S_3^1 S_3^5R	S_3^5 S_3^1R	C_3^1 C_3^2R	C_3^2 C_3^1R	σ_h σ_hR	$3C'_2$ $3C'_2R$	$3\sigma_v$ $3\sigma_vR$
C_{6v}		E	R	C_6 C_6^5R	C_6^5 C_6R	C_3 C_3^2R	C_3^2 C_3R	C_2 C_2R	$3\sigma_v$ $3\sigma_vR$	$3\sigma_d$ $3\sigma_dR$
D_6 (h = 24)		E	R	C_6 C_6^5R	C_6^5 C_6R	C_3 C_3^2R	C_3^2 C_3R	C_2 C_2R	$3C'_2$ $3C'_2R$	$3C''_2$ $3C''_2R$
Γ_1	A_1	+1	+1	+1	+1	+1	+1	+1	+1	+1
Γ_2	A_2	+1	+1	+1	+1	+1	+1	+1	-1	-1
Γ_3	B_1	+1	+1	-1	-1	+1	+1	-1	+1	-1
Γ_4	B_2	+1	+1	-1	-1	+1	+1	-1	-1	+1
Γ_5	E_1	+2	+2	+1	+1	-1	-1	-2	0	0
Γ_6	E_2	+2	+2	-1	-1	-1	-1	+2	0	0
Γ_7	$E_{\frac{1}{2}}$	+2	-2	$+\sqrt{3}$	$-\sqrt{3}$	+1	-1	0	0	0
Γ_8	$E_{\frac{3}{2}}$	+2	-2	0	0	-2	+2	0	0	0
Γ_9	$E_{\frac{5}{2}}$	+2	-2	$-\sqrt{3}$	$+\sqrt{3}$	+1	-1	0	0	0

D_{2d}		E	R	S_4 S_4^3R	S_4^3 S_4R	C_2 C_2R	$2C'_2$ $2C'_2R$	$2\sigma_d$ $2\sigma_dR$
C_{4v}		E	R	C_4 C_4^3R	C_4^3 C_4R	C_2 C_2R	$2\sigma_v$ $2\sigma_vR$	$2\sigma_d$ $2\sigma_dR$
D_4 (h = 16)		E	R	C_4 C_4^3R	C_4^3 C_4R	C_2 C_2R	$2C'_2$ $2C'_2R$	$2C''_2$ $2C''_2R$
Γ_1	A_1	+1	+1	+1	+1	+1	+1	+1
Γ_2	A_2	+1	+1	+1	+1	+1	-1	-1
Γ_3	B_1	+1	+1	-1	-1	+1	+1	-1
Γ_4	B_2	+1	+1	-1	-1	+1	-1	+1
Γ_5	E_1	+2	+2	0	0	-2	0	0
Γ_6	$E_{\frac{1}{2}}$	+2	-2	$+\sqrt{2}$	$-\sqrt{2}$	0	0	0
Γ_7	$E_{\frac{3}{2}}$	+2	-2	$-\sqrt{2}$	$+\sqrt{2}$	0	0	0

C_{3v}		E	R	C_3, C_3^2R	C_3^2, C_3R	$3\sigma_v$	$3\sigma_vR$
D_3 (h = 12)		E	R	C_3, C_3^2R	C_3^2, C_3R	$3C_2$	$3C_2R$
Γ_1	A_1	+1	+1	+1	+1	+1	+1
Γ_2	A_2	+1	+1	+1	+1	-1	-1
Γ_3	E_1	+2	+2	-1	-1	0	0
Γ_4	$E_{\frac{1}{2}}$	+2	-2	+1	-1	0	0
Γ_5 }	$E_{\frac{3}{2}}$	{ +1	-1	-1	+1	+i	-i }
Γ_6 }		{ +1	-1	-1	+1	-i	+i }

C_{2v}		E	R	C_2, C_2R	σ_v, σ_vR	σ_d, σ_dR
D_2 (h = 8)		E	R	$C_2(z)$, $C_2(z)R$	$C_2(y)$, $C_2(y)R$	$C_2(x)$, $C_2(x)R$
Γ_1	A_1	+1	+1	+1	+1	+1
Γ_2	B_1	+1	+1	+1	-1	-1
Γ_3	B_2	+1	+1	-1	+1	-1
Γ_4	B_3	+1	+1	-1	-1	+1
Γ_5	$E_{\frac{1}{2}}$	+2	-2	0	0	0

Some characters of the double pure rotation group, **R(3)'**. The upper left hand quadrant of this table contains the characters of the pure rotation group in three dimensions, **R(3)**.

R(3)'		E	$C_2(C_2R)$	C_3	C_4	C_5	C_6	...	R	C_3R	C_4R	C_5R	C_6R
$D^{(0)}$	S	+1	+1	+1	+1	+1	+1	...	+1	+1	+1	+1	+1
$D^{(1)}$	P	+3	-1	0	+1	$+\tau$	+2	...	+3	0	+1	$+\tau$	+2
$D^{(2)}$	D	+5	+1	-1	-1	0	+1	...	+5	-1	-1	0	+1
$D^{(3)}$	F	+7	-1	+1	-1	$-\tau$	-1	...	+7	+1	-1	$-\tau$	-1
$D^{(4)}$	G	+9	+1	0	+1	-1	-2	...	+9	0	+1	-1	-2
$D^{(5)}$	H	+11	-1	-1	+1	+1	-1	...	+11	-1	+1	+1	-1
$D^{(6)}$	I	+13	+1	+1	-1	$+\tau$	+1	...	+13	+1	-1	$+\tau$	+1
⋮													
$D^{(\frac{1}{2})}$		+2	0	+1	$+\sqrt{2}$	$+\tau$	$+\sqrt{3}$	...	-2	-1	$-\sqrt{2}$	$-\tau$	$-\sqrt{3}$
$D^{(3/2)}$		+4	0	-1	0	+1	$+\sqrt{3}$	...	-4	+1	0	-1	$-\sqrt{3}$
$D^{(5/2)}$		+6	0	0	$-\sqrt{2}$	-1	0	...	-6	0	$+\sqrt{2}$	+1	0
$D^{(7/2)}$		+8	0	+1	0	$-\tau$	$-\sqrt{3}$	...	-8	-1	0	$+\tau$	$+\sqrt{3}$
$D^{(9/2)}$		+10	0	-1	$+\sqrt{2}$	0	$-\sqrt{3}$	...	-10	+1	$-\sqrt{2}$	0	$+\sqrt{3}$
$D^{(11/2)}$		+12	0	0	0	$+\tau$	0	...	-12	0	0	$-\tau$	0
⋮													

7 Group correlations

7.1 The hierarchy of point groups and their subgroups

The common dihedral and cyclic point groups have been arranged in a table in descending sequence from groups of highest order to those of lowest. This enables all the subgroups of a particular point group to be noted at a glance. The quotient of the order of a group divided by that of any of its subgroups must be a small integer greater than unity. This relates to the table in so far as the immediate subgroups of any point group are to be found in the row below the point group in question.

The icosahedral and cubic groups have been arranged in a similar table and the descent in symmetry continued until correspondence with the other table is achieved.

The correlation of the three-dimensional pure rotation, and rotation-inversion groups, **R**(3) and **O**(3), with point groups of lower symmetry may be found in §8. 6.

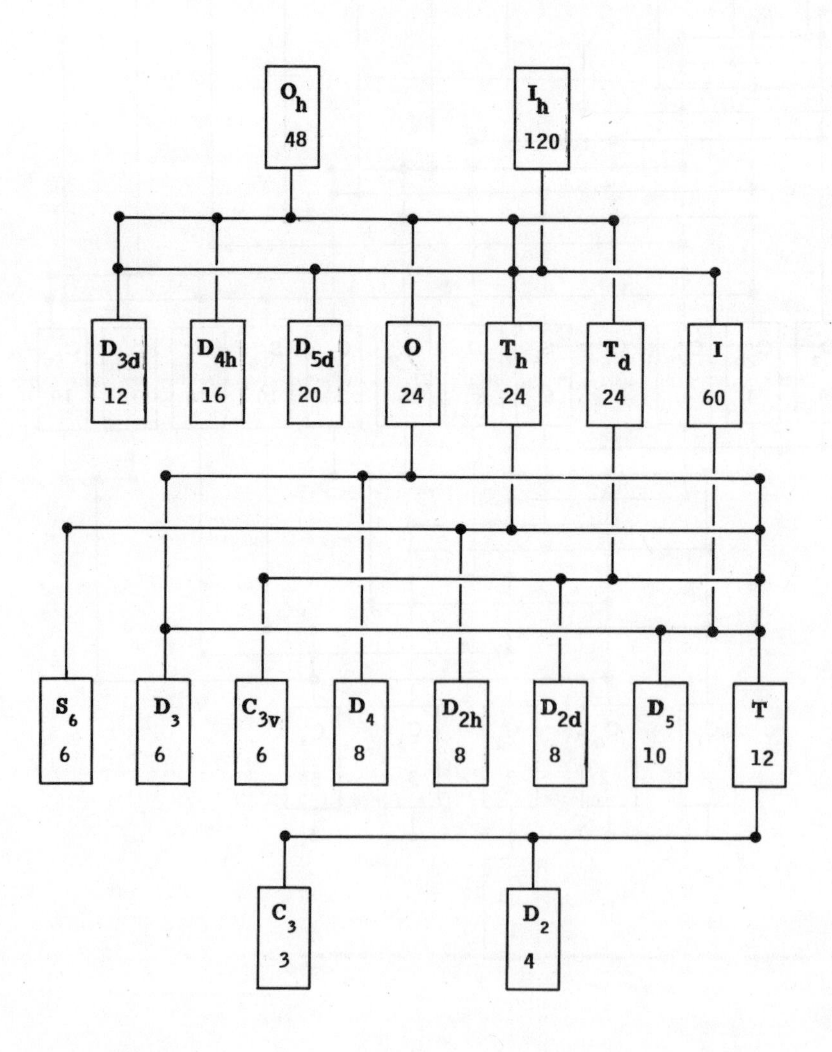

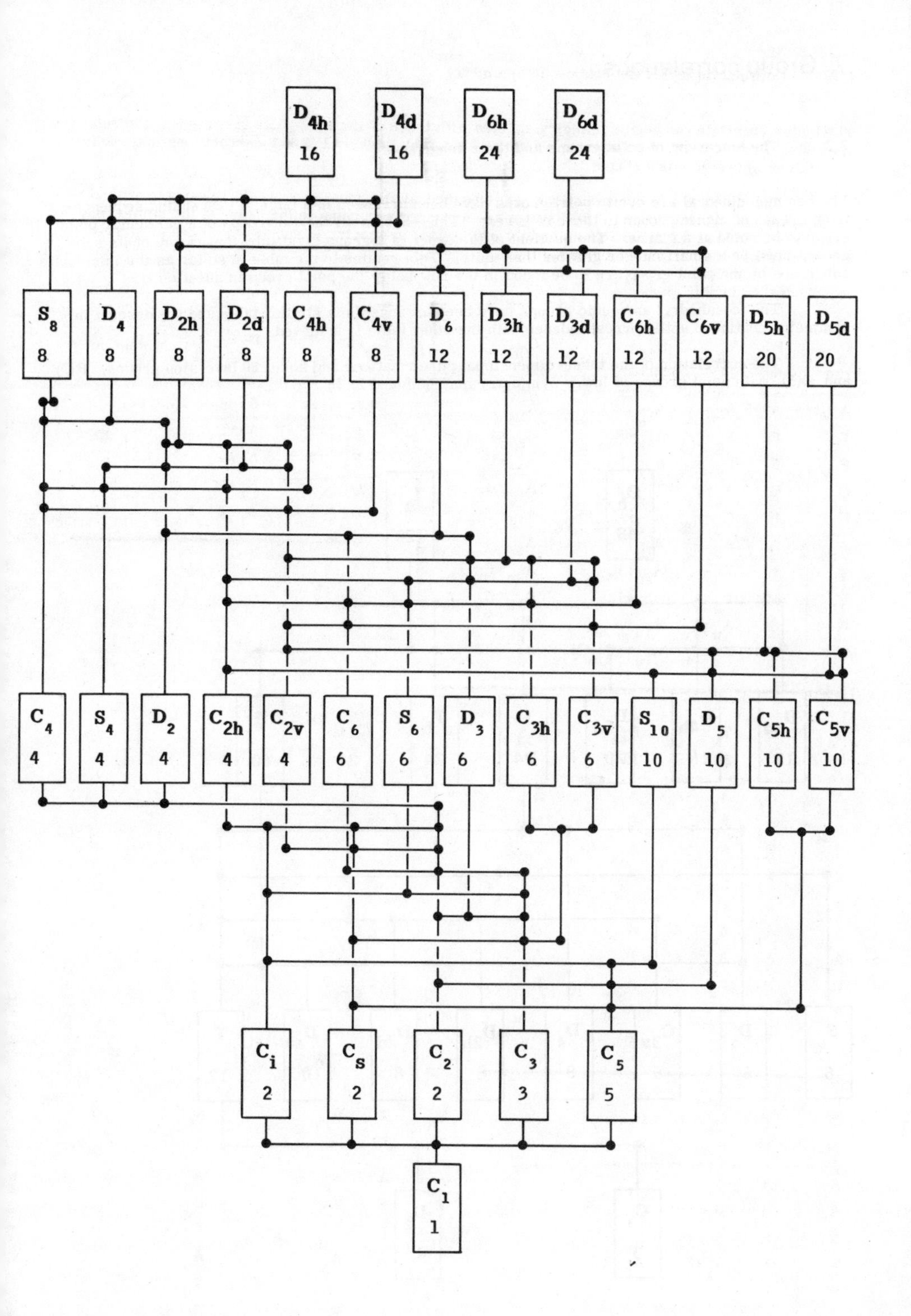
D4h 16
D4d 16
D6h 24
D6d 24
S8 8
D4 8
D2h 8
D2d 8
C4h 8
C4v 8
D6 12
D3h 12
D3d 12
C6h 12
C6v 12
D5h 20
D5d 20
C4 4
S4 4
D2 4
C2h 4
C2v 4
C6 6
S6 6
D3 6
C3h 6
C3v 6
S10 10
D5 10
C5h 10
C5v 10
Ci 2
Cs 2
C2 2
C3 3
C5 5
C1 1

7.2 Correlation tables for descent in symmetry

The tables correlate the point symmetry species with those of the immediate subgroups. Correlations are listed directly between adjacent rows in the hierarchy tables. Further correlations may easily be obtained in two or more stages.

The tables are in three sections, (a) icosahedral and cubic groups, (b) linear groups and (c) dihedral and cyclic groups. The listings are arranged according to the order of the point group - from highest to lowest. Where an ambiguity may arise, symmetry elements which are preserved are noted above the particular subgroup listing.

Icosahedral and cubic groups

I_h 120	I 60	T_h 24	D_{5d} 20	D_{3d} 12
A_g	A	A_g	A_{1g}	A_{1g}
F_{1g}	F_1	F_g	$A_{2g}+E_{1g}$	$A_{2g}+E_g$
F_{2g}	F_2	F_g	$A_{2g}+E_{2g}$	$A_{2g}+E_g$
G_g	G	A_g+F_g	$E_{1g}+E_{2g}$	$A_{1g}+A_{2g}+E_g$
H_g	H	E_g+F_g	$A_{1g}+E_{1g}+E_{2g}$	$A_{1g}+2E_g$
A_u	A	A_u	A_{1u}	A_{1u}
F_{1u}	F_1	F_u	$A_{2u}+E_{1u}$	$A_{2u}+E_u$
F_{2u}	F_2	F_u	$A_{2u}+E_{2u}$	$A_{2u}+E_u$
G_u	G	A_u+F_u	$E_{1u}+E_{2u}$	$A_{1u}+A_{2u}+E_u$
H_u	H	E_u+F_u	$A_{1u}+E_{1u}+E_{2u}$	$A_{1u}+2E_u$

I 60	T 12	D_5 10	D_3 6
A	A	A_1	A_1
F_1	F	A_2+E_1	A_2+E
F_2	F	A_2+E_2	A_2+E
G	$A+F$	E_1+E_2	A_1+A_2+E
H	$E+F$	$A_1+E_1+E_2$	A_1+2E

O_h 48	O 24	T_h 24	T_d 24	D_{4h} 16	D_{3d} 12
A_{1g}	A_1	A_g	A_1	A_{1g}	A_{1g}
A_{2g}	A_2	A_g	A_2	B_{1g}	A_{2g}
E_g	E	E_g	E	$A_{1g}+B_{1g}$	E_g
F_{1g}	F_1	F_g	F_1	$A_{2g}+E_g$	$A_{2g}+E_g$
F_{2g}	F_2	F_g	F_2	$B_{2g}+E_g$	$A_{1g}+E_g$
A_{1u}	A_1	A_u	A_2	A_{1u}	A_{1u}
A_{2u}	A_2	A_u	A_1	B_{1u}	A_{2u}
E_u	E	E_u	E	$A_{1u}+B_{1u}$	E_u
F_{1u}	F_1	F_u	F_2	$A_{2u}+E_u$	$A_{2u}+E_u$
F_{2u}	F_2	F_u	F_1	$B_{2u}+E_u$	$A_{1u}+E_u$

O 24	T 12	D_4 8	D_3 6
A_1	A	A_1	A_1
A_2	A	B_1	A_2
E	E	A_1+B_1	E
F_1	F	A_2+E	A_2+E
F_2	F	B_2+E	A_1+E

T_h 24	T 12	D_{2h} 8	S_6 6
A_g	A	A_g	A_g
E_g	E	$2A_g$	E_g
F_g	F	$B_{1g}+B_{2g}+B_{3g}$	A_g+E_g
A_u	A	A_u	A_u
E_u	E	$2A_u$	E_u
F_u	F	$B_{1u}+B_{2u}+B_{3u}$	A_u+E_u

T_d 24	T 12	D_{2d} 8	C_{3v} 6
A_1	A	A_1	A_1
A_2	A	B_1	A_2
E	E	A_1+B_1	E
F_1	F	A_2+E	A_2+E
F_2	F	B_2+E	A_1+E

T 12	D_2 4	C_3 3
A	A	A
E	$2A$	E
F	$B_1+B_2+B_3$	$A+E$

'Linear groups'

$C_{\infty v}$ ∞	C_∞ ∞	C_{8v} 16	C_{7v} 14	C_{6v} 12	C_{5v} 10	C_{4v} 8
Σ^+	Σ	A_1	A_1	A_1	A_1	A_1
Σ^-	Σ	A_2	A_2	A_2	A_2	A_2
Π	Π	E_1	E_1	E_1	E_1	E
Δ	Δ	E_2	E_2	E_2	E_2	B_1+B_2
Φ	Φ	E_3	E_3	B_1+B_2	E_2	E
Γ	Γ	B_1+B_2	E_3	E_2	E_1	A_1+A_2

$D_{\infty h}$ ∞	D_∞ ∞	D_{6h} 24	D_{5h} 20	D_{4h} 16	D_{6d} 24	D_{5d} 20	D_{4d} 16
Σ_g^+	Σ^+	A_{1g}	A'_1	A_{1g}	A_1	A_{1g}	A_1
Σ_u^+	Σ^+	A_{2u}	A''_2	A_{2u}	B_2	A_{2u}	B_2
Σ_g^-	Σ^-	A_{2g}	A'_2	A_{2g}	A_2	A_{2g}	A_2
Σ_u^-	Σ^-	A_{1u}	A''_1	A_{1u}	B_1	A_{1u}	B_1
Π_g	Π	E_{1g}	E''_1	E_g	E_5	E_{1g}	E_3
Π_u	Π	E_{1u}	E'_1	E_u	E_1	E_{1u}	E_1
Δ_g	Δ	E_{2g}	E'_2	$B_{1g}+B_{2g}$	E_2	E_{2g}	E_2
Δ_u	Δ	E_{2u}	E''_2	$B_{1u}+B_{2u}$	E_4	E_{2u}	E_2
Φ_g	Φ	$B_{1g}+B_{2g}$	E''_2	E_g	E_3	E_{2g}	E_1
Φ_u	Φ	$B_{1u}+B_{2u}$	E'_2	E_u	E_3	E_{2u}	E_3
Γ_g	Γ	E_{2g}	E'_1	$A_{1g}+A_{2g}$	E_4	E_{1g}	B_1+B_2
Γ_u	Γ	E_{2u}	E''_1	$A_{1u}+A_{2u}$	E_2	E_{1u}	A_1+A_2

Axial groups

D_{6h} 24	D_6 12	C'_2 D_{3h} 12	C''_2 D_{3h} 12	C'_2 D_{3d} 12	C''_2 D_{3d} 12	C_{6h} 12	C_{6v} 12	$\sigma_v \rightarrow \sigma(yz)$ $\sigma_h \rightarrow \sigma(xy)$ D_{2h} 8
A_{1g}	A_1	A'_1	A'_1	A_{1g}	A_{1g}	A_g	A_1	A_g
A_{2g}	A_2	A'_2	A'_2	A_{2g}	A_{2g}	A_g	A_2	B_{1g}
B_{1g}	B_1	A''_1	A''_2	A_{1g}	A_{2g}	B_g	B_2	B_{2g}
B_{2g}	B_2	A''_2	A''_1	A_{2g}	A_{1g}	B_g	B_1	B_{3g}
E_{1g}	E_1	E''	E''	E_g	E_g	E_{1g}	E_1	$B_{2g}+B_{3g}$
E_{2g}	E_2	E'	E'	E_g	E_g	E_{2g}	E_2	A_g+B_{1g}
A_{1u}	A_1	A''_1	A''_1	A_{1u}	A_{1u}	A_u	A_2	A_u
A_{2u}	A_2	A''_2	A''_2	A_{2u}	A_{2u}	A_u	A_1	B_{1u}
B_{1u}	B_1	A'_1	A'_2	A_{1u}	A_{2u}	B_u	B_1	B_{2u}
B_{2u}	B_2	A'_2	A'_1	A_{2u}	A_{1u}	B_u	B_2	B_{3u}
E_{1u}	E_1	E'	E'	E_u	E_u	E_{1u}	E_1	$B_{2u}+B_{3u}$
E_{2u}	E_2	E''	E''	E_u	E_u	E_{2u}	E_2	A_u+B_{1u}

D_{6d} 24	D_6 12	C_{6v} 12	D_{2d} 8
A_1	A_1	A_1	A_1
A_2	A_2	A_2	A_2
B_1	A_1	A_2	B_1
B_2	A_2	A_1	B_2
E_1	E_1	E_1	E
E_2	E_2	E_2	B_1+B_2
E_3	B_1+B_2	B_1+B_2	E
E_4	E_2	E_2	A_1+A_2
E_5	E_1	E_1	E

D_{5h} 20	D_5 10	C_{5h} 10	C_{5v} 10	$\sigma_h \rightarrow \sigma(xz)$ C_{2v} 4
A'_1	A_1	A'	A_1	A_1
A'_2	A_2	A'	A_2	B_1
E'_1	E_1	E'_1	E_1	A_1+B_1
E'_2	E_2	E'_2	E_2	A_1+B_1
A''_1	A_1	A''	A_2	A_2
A''_2	A_2	A''	A_1	B_2
E''_1	E_1	E''_1	E_1	A_2+B_2
E''_2	E_2	E''_2	E_2	A_2+B_2

D_{5d} 20	S_{10} 10	D_5 10	C_{5v} 10	C_{2h} 4
A_{1g}	A_g	A_1	A_1	A_g
A_{2g}	A_g	A_2	A_2	B_g
E_{1g}	E_{1g}	E_1	E_1	A_g+B_g
E_{2g}	E_{2g}	E_2	E_2	A_g+B_g
A_{1u}	A_u	A_1	A_2	A_u
A_{2u}	A_u	A_2	A_1	B_u
E_{1u}	E_{1u}	E_1	E_1	A_u+B_u
E_{2u}	E_{2u}	E_2	E_2	A_u+B_u

D_{4h} 16	D_4 8	C'_2 D_{2h} 8	C''_2 D_{2h} 8	C'_2 D_{2d} 8	C''_2 D_{2d} 8	C_{4h} 8	C_{4v} 8
A_{1g}	A_1	A_g	A_g	A_1	A_1	A_g	A_1
A_{2g}	A_2	B_{1g}	B_{1g}	A_2	A_2	A_g	A_2
B_{1g}	B_1	A_g	B_{1g}	B_1	B_2	B_g	B_1
B_{2g}	B_2	B_{1g}	A_g	B_2	B_1	B_g	B_2
E_g	E	$B_{2g}+B_{3g}$	$B_{2g}+B_{3g}$	E	E	E_g	E
A_{1u}	A_1	A_u	A_u	B_1	B_1	A_u	A_2
A_{2u}	A_2	B_{1u}	B_{1u}	B_2	B_2	A_u	A_1
B_{1u}	B_1	A_u	B_{1u}	A_1	A_2	B_u	B_2
B_{2u}	B_2	B_{1u}	A_u	A_2	A_1	B_u	B_1
E_u	E	$B_{2u}+B_{3u}$	$B_{2u}+B_{3u}$	E	E	E_u	E

D_{4d} 16	S_8 8	D_4 8	C_{4v} 8
A_1	A	A_1	A_1
A_2	A	A_2	A_2
B_1	B	A_1	A_2
B_2	B	A_2	A_1
E_1	E_1	E	E
E_2	E_2	B_1+B_2	B_1+B_2
E_3	E_3	E	E

D_6 12	C_6 6	C'_2 D_3 6	C''_2 D_3 6	D_2 4
A_1	A	A_1	A_1	A
A_2	A	A_2	A_2	B_1
B_1	B	A_1	A_2	B_2
B_2	B	A_2	A_1	B_3
E_1	E_1	E	E	B_2+B_3
E_2	E_2	E	E	$A+B_1$

D_{3h} 12	D_3 6	C_{3h} 6	C_{3v} 6	$\sigma_h \to \sigma(yz)$ C_{2v} 4
A'_1	A_1	A'	A_1	A_1
A'_2	A_2	A'	A_2	B_2
E'	E	E'	E	A_1+B_2
A''_1	A_1	A"	A_2	A_2
A''_2	A_2	A"	A_1	B_1
E"	E	E"	E	A_2+B_1

D_{3d} 12	S_6 6	D_3 6	C_{3v} 6	C_{2h} 4
A_{1g}	A_g	A_1	A_1	A_g
A_{2g}	A_g	A_2	A_2	B_g
E_g	E_g	E	E	A_g+B_g
A_{1u}	A_u	A_1	A_2	A_u
A_{2u}	A_u	A_2	A_1	B_u
E_u	E_u	E	E	A_u+B_u

C_{6h} 12	C_6 6	S_6 6	C_{3h} 6	C_{2h} 4
A_g	A	A_g	A'	A_g
B_g	B	A_g	A"	B_g
E_{1g}	E_1	E_g	E"	$2B_g$
E_{2g}	E_2	E_g	E'	$2A_g$
A_u	A	A_u	A"	A_u
B_u	B	A_u	A'	B_u
E_{1u}	E_1	E_u	E'	$2B_u$
E_{2u}	E_2	E_u	E"	$2A_u$

C_{6v} 12	C_6 6	σ_v C_{3v} 6	σ_d C_{3v} 6	$\sigma_v \to \sigma(xz)$ C_{2v} 4
A_1	A	A_1	A_1	A_1
A_2	A	A_2	A_2	A_2
B_1	B	A_1	A_2	B_1
B_2	B	A_2	A_1	B_2
E_1	E_1	E	E	B_1+B_2
E_2	E_2	E	E	A_1+A_2

S_{10} 10	C_5 5	C_i 2
A_g	A	A_g
E_{1g}	E_1	$2A_g$
E_{2g}	E_2	$2A_g$
A_u	A	A_u
E_{1u}	E_1	$2A_u$
E_{2u}	E_2	$2A_u$

D_5 10	C_5 5	C_2 2
A_1	A	A
A_2	A	B
E_1	E_1	A+B
E_2	E_2	A+B

C_{5h} 10	C_5 5	C_s 2
A'	A	A'
E'_1	E_1	2A'
E'_2	E_2	2A'
A"	A	A"
E''_1	E_1	2A"
E''_2	E_2	2A"

C_{5v} 10	C_5 5	C_s 2
A_1	A	A'
A_2	A	A"
E_1	E_1	A'+A"
E_2	E_2	A'+A"

S_8	C_4
8	4
A	A
B	A
E_1	E
E_2	2B
E_3	E

C_8	C_4
8	4
A	A
B	A
E_1	E
E_2	2B
E_3	E

		C'_2	C''_2
D_4	C_4	D_2	D_2
8	4	4	4
A_1	A	A	A
A_2	A	B_1	B_1
B_1	B	A	B_1
B_2	B	B_1	A
E	E	B_2+B_3	B_2+B_3

		$C_2(z)$	$C_2(y)$	$C_2(x)$	$C_2(z)$	$C_2(y)$	$C_2(x)$
D_{2h}	D_2	C_{2h}	C_{2h}	C_{2h}	C_{2v}	C_{2v}	C_{2v}
8	4	4	4	4	4	4	4
A_g	A	A_g	A_g	A_g	A_1	A_1	A_1
B_{1g}	B_1	A_g	B_g	B_g	A_2	B_2	B_1
B_{2g}	B_2	B_g	A_g	B_g	B_1	A_2	B_2
B_{3g}	B_3	B_g	B_g	A_g	B_2	B_1	A_2
A_u	A	A_u	A_u	A_u	A_2	A_2	A_2
B_{1u}	B_1	A_u	B_u	B_u	A_1	B_1	B_2
B_{2u}	B_2	B_u	A_u	B_u	B_2	A_1	B_1
B_{3u}	B_3	B_u	B_u	A_u	B_1	B_2	A_1

		$C_2 \to C_2(z)$	
D_{2d}	S_4	D_2	C_{2v}
8	4	4	4
A_1	A	A	A_1
A_2	A	B_1	A_2
B_1	B	A	A_2
B_2	B	B_1	A_1
E	E	B_2+B_3	B_1+B_2

C_{4h}	C_4	S_4	C_{2h}
8	4	4	4
A_g	A	A	A_g
B_g	B	B	A_g
E_g	E	E	$2B_g$
A_u	A	B	A_u
B_u	B	A	A_u
E_u	E	E	$2B_u$

		σ_v	σ_d
C_{4v}	C_4	C_{2v}	C_{2v}
8	4	4	4
A_1	A	A_1	A_1
A_2	A	A_2	A_2
B_1	B	A_1	A_2
B_2	B	A_2	A_1
E	E	B_1+B_2	B_1+B_2

C_6	C_3	C_2
6	3	2
A	A	A
B	A	B
E_1	E	2B
E_2	E	2A

S_6	C_3	C_i
6	3	2
A_g	A	A_g
E_g	E	$2A_g$
A_u	A	A_u
E_u	E	$2A_u$

D_3	C_3	C_2
6	3	2
A_1	A	A
A_2	A	B
E	E	A+B

C_{3h}	C_3	C_s
6	3	2
A'	A	A'
E'	E	2A'
A"	A	A"
E"	E	2A"

C_{3v} 6	C_3 3	C_s 2
A_1	A	A'
A_2	A	A"
E	E	A'+A"

C_4 4	C_2 2
A	A
B	A
E	2B

S_4 4	C_2 2
A	A
B	A
E	2B

D_2 4	C_2(z) 2	C_2(y) 2	C_2(x) 2
A	A	A	A
B_1	A	B	B
B_2	B	A	B
B_3	B	B	A

C_{2h} 4	C_2 2	C_s 2	C_i 2
A_g	A	A'	A_g
B_g	B	A"	A_g
A_u	A	A"	A_u
B_u	B	A'	A_u

C_{2v} 4	C_2 2	σ (xz) C_s 2	σ (yz) C_s 2
A_1	A	A'	A'
A_2	A	A"	A"
B_1	B	A'	A"
B_2	B	A"	A'

7.3 Site symmetries in the space groups

The following table is intended for use with the Halford Site Symmetry method for analysis of the vibrational spectra of solids,[10] and lists all the space groups having higher site symmetries than C_1. The space group[11] is recorded as (i) its number, as found in the 'International Tables for X-ray Crystallography' vol. I, (ii) its Hermann-Mauguin symbol and (iii) its Schönflies symbol. The site symmetry listings describe all the sites present. The standard Schönflies point group symbol is preceded by the number of distinct sets of sites of that symmetry and followed in parentheses by the number of equivalent sites per set. In the case of point groups with one or more degrees of translational freedom in the lattice (C_n, C_{nv} and C_s) any number of distinct atoms may occupy these sites. The number of equivalent sites per set is given for the conventional unit cell and should be divided by the cell multiplicity to obtain the factor group correlation number (i. e. the equivalent sites per primitive cell). The various cell multiplicities are:

4 for an F cell,

2 for I, A and C cells,

1 for R and P cells.

Space group		Site symmetries
2 P$\bar{1}$	C_i^1	8C_i
3 P2	C_2^1	4C_2
5 C2	C_2^3	2C_2(2)
6 Pm	C_s^1	2C_s
8 Cm	C_s^3	C_s(2)
10 P2/m	C_{2h}^1	8C_{2h}; 4C_2(2); 2C_s(2)
11 P2_1/m	C_{2h}^2	4C_i(2); C_s(2)
12 C2/m	C_{2h}^3	4C_{2h}(2); 2C_i(4); 2C_2(4); C_s(4)

Space group		Site symmetries
13 P2/c	C_{2h}^{4}	$4C_i(2)$; $2C_2(2)$
14 $P2_1/c$	C_{2h}^{5}	$4C_i(2)$
15 C2/c	C_{2h}^{6}	$4C_i(4)$; $C_2(4)$
16 P222	D_{2}^{1}	$8D_2$; $12C_2(2)$
17 $P222_1$	D_{2}^{2}	$4C_2(2)$
18 $P2_12_12$	D_{2}^{3}	$2C_2(2)$
20 $C222_1$	D_{2}^{5}	$2C_2(4)$
21 C222	D_{2}^{6}	$4D_2(2)$; $7C_2(4)$
22 F222	D_{2}^{7}	$4D_2(4)$; $6C_2(8)$
23 I222	D_{2}^{8}	$4D_2(2)$; $6C_2(4)$
24 $I2_12_12_1$	D_{2}^{9}	$3C_2(4)$
25 Pmm2	C_{2v}^{1}	$4C_{2v}$; $4C_s(2)$
26 $Pmc2_1$	C_{2v}^{2}	$2C_s(2)$
27 Pcc2	C_{2v}^{3}	$4C_2(2)$
28 Pma2	C_{2v}^{4}	$2C_2(2)$; $C_s(2)$
30 Pnc2	C_{2v}^{6}	$2C_2(2)$
31 $Pmn2_1$	C_{2v}^{7}	$C_s(2)$
32 Pba2	C_{2v}^{8}	$2C_2(2)$
34 Pnn2	C_{2v}^{10}	$2C_2(2)$
35 Cmm2	C_{2v}^{11}	$2C_{2v}(2)$; $C_2(4)$; $2C_s(4)$
36 $Cmc2_1$	C_{2v}^{12}	$C_s(4)$
37 Ccc2	C_{2v}^{13}	$3C_2(4)$
38 Amm2	C_{2v}^{14}	$2C_{2v}(2)$; $3C_s(4)$
39 Abm2	C_{2v}^{15}	$2C_2(4)$; $C_s(4)$
40 Ama2	C_{2v}^{16}	$C_2(4)$; $C_s(4)$
41 Aba2	C_{2v}^{17}	$C_2(4)$
42 Fmm2	C_{2v}^{18}	$C_{2v}(4)$; $C_2(8)$; $2C_s(8)$
43 Fdd2	C_{2v}^{19}	$C_2(8)$
44 Imm2	C_{2v}^{20}	$2C_{2v}(2)$; $2C_s(4)$
45 Iba2	C_{2v}^{21}	$2C_2(4)$
46 Ima2	C_{2v}^{22}	$C_2(4)$; $C_s(4)$
47 Pmmm	D_{2h}^{1}	$8D_{2h}$; $12C_{2v}(2)$; $6C_s(4)$
48 Pnnn	D_{2h}^{2}	$4D_2(2)$; $2C_i(4)$; $6C_2(4)$

Space group			Site symmetries
49	Pccm	D_{2h}^{3}	$4C_{2h}(2)$; $4D_2(2)$; $8C_2(4)$; $C_s(4)$
50	Pban	D_{2h}^{4}	$4D_2(2)$; $2C_i(4)$; $6C_2(4)$
51	Pmma	D_{2h}^{5}	$4C_{2h}(2)$; $2C_{2v}(2)$; $2C_2(4)$; $3C_s(4)$
52	Pnna	D_{2h}^{6}	$2C_i(4)$; $2C_2(4)$
53	Pmna	D_{2h}^{7}	$4C_{2h}(2)$; $3C_2(4)$; $C_s(4)$
54	Pcca	D_{2h}^{8}	$2C_i(4)$; $3C_2(4)$
55	Pbam	D_{2h}^{9}	$4C_{2h}(2)$; $2C_2(4)$; $2C_s(4)$
56	Pccn	D_{2h}^{10}	$2C_i(4)$; $2C_2(4)$
57	Pbcm	D_{2h}^{11}	$2C_i(4)$; $C_2(4)$; $C_s(4)$
58	Pnnm	D_{2h}^{12}	$4C_{2h}(2)$; $2C_2(4)$; $C_s(4)$
59	Pmmn	D_{2h}^{13}	$2C_{2v}(2)$; $2C_i(4)$; $2C_s(4)$
60	Pbcn	D_{2h}^{14}	$2C_i(4)$; $C_2(4)$
61	Pbca	D_{2h}^{15}	$2C_i(4)$
62	Pnma	D_{2h}^{16}	$2C_i(4)$; $C_s(4)$
63	Cmcm	D_{2h}^{17}	$2C_{2h}(4)$; $C_{2v}(4)$; $C_i(8)$; $C_2(8)$; $2C_s(8)$
64	Cmca	D_{2h}^{18}	$2C_{2h}(4)$; $C_i(8)$; $C_s(8)$; $2C_2(8)$
65	Cmmm	D_{2h}^{19}	$4D_{2h}(2)$; $2C_{2h}(4)$; $6C_{2v}(4)$; $C_2(8)$; $4C_s(8)$
66	Cccm	D_{2h}^{20}	$2D_2(4)$; $4C_{2h}(4)$; $5C_2(8)$; $C_s(8)$
67	Cmma	D_{2h}^{21}	$2D_2(4)$; $4C_{2h}(4)$; $C_{2v}(4)$; $5C_2(8)$; $2C_s(8)$
68	Ccca	D_{2h}^{22}	$2D_2(4)$; $2C_i(8)$; $4C_2(8)$
69	Fmmm	D_{2h}^{23}	$2D_{2h}(4)$; $3C_{2h}(8)$; $D_2(8)$; $3C_{2v}(8)$; $3C_2(16)$; $3C_s(16)$
70	Fddd	D_{2h}^{24}	$2D_2(8)$; $2C_i(16)$; $3C_2(16)$
71	Immm	D_{2h}^{25}	$4D_{2h}(2)$; $6C_{2v}(4)$; $C_i(8)$; $3C_s(8)$
72	Ibam	D_{2h}^{26}	$2D_2(4)$; $2C_{2h}(4)$; $C_i(8)$; $4C_2(8)$; $C_s(8)$
73	Ibca	D_{2h}^{27}	$2C_i(8)$; $3C_2(8)$
74	Imma	D_{2h}^{28}	$4C_{2h}(4)$; $C_{2v}(4)$; $2C_2(8)$; $2C_s(8)$
75	P4	C_4^1	$2C_4$; $C_2(2)$
77	$P4_2$	C_4^3	$3C_2(2)$
79	I4	C_4^5	$C_4(2)$; $C_2(4)$
80	$I4_1$	C_4^6	$C_2(4)$
81	$P\bar{4}$	S_4^1	$4S_4$; $3C_2(2)$
82	$I\bar{4}$	S_4^2	$4S_4(2)$; $2C_2(4)$
83	P4/m	C_{4h}^1	$4C_{4h}$; $2C_{2h}(2)$; $2C_4(2)$; $C_2(4)$; $2C_s(4)$
84	$P4_2/m$	C_{4h}^2	$4C_{2h}(2)$; $2S_4(2)$; $3C_2(4)$; $C_s(4)$

Space group			Site symmetries
85	$P4/n$	C_{4h}^3	$2S_4(2)$; $C_4(2)$; $2C_i(4)$; $C_2(4)$
86	$P4_2/n$	C_{4h}^4	$2S_4(2)$; $2C_i(4)$; $2C_2(4)$
87	$I4/m$	C_{4h}^5	$2C_{4h}(2)$; $C_{2h}(4)$; $S_4(4)$; $C_4(4)$; $C_i(8)$; $C_2(8)$; $C_s(8)$
88	$I4_1/a$	C_{4h}^6	$2S_4(4)$; $2C_i(8)$; $C_2(8)$
89	$P422$	D_4^1	$4D_4$; $2D_2(2)$; $2C_4(2)$; $7C_2(4)$
90	$P42_12$	D_4^2	$2D_2(2)$; $C_4(2)$; $3C_2(4)$
91	$P4_122$	D_4^3	$3C_2(4)$
92	$P4_12_12$	D_4^4	$C_2(4)$
93	$P4_222$	D_4^5	$6D_2(2)$; $9C_2(4)$
94	$P4_22_12$	D_4^6	$2D_2(2)$; $4C_2(4)$
95	$P4_322$	D_4^7	$3C_2(4)$
96	$P4_32_12$	D_4^8	$C_2(4)$
97	$I422$	D_4^9	$2D_4(2)$; $2D_2(4)$; $C_4(4)$; $5C_2(8)$
98	$I4_122$	D_4^{10}	$2D_2(4)$; $4C_2(8)$
99	$P4mm$	C_{4v}^1	$2C_{4v}$; $C_{2v}(2)$; $3C_s(4)$
100	$P4bm$	C_{4v}^2	$C_4(2)$; $C_{2v}(2)$; $C_s(4)$
101	$P4_2cm$	C_{4v}^3	$2C_{2v}(2)$; $C_2(4)$; $C_s(4)$
102	$P4_2nm$	C_{4v}^4	$C_{2v}(2)$; $C_2(4)$; $C_s(4)$
103	$P4cc$	C_{4v}^5	$2C_4(2)$; $C_2(4)$
104	$P4nc$	C_{4v}^6	$C_4(2)$; $C_2(4)$
105	$P4_2mc$	C_{4v}^7	$3C_{2v}(2)$; $2C_s(4)$
106	$P4_2bc$	C_{4v}^8	$2C_2(4)$
107	$I4mm$	C_{4v}^9	$C_{4v}(2)$; $C_{2v}(4)$; $2C_s(8)$
108	$I4cm$	C_{4v}^{10}	$C_4(4)$; $C_{2v}(4)$; $C_s(8)$
109	$I4_1md$	C_{4v}^{11}	$C_{2v}(4)$; $C_s(8)$
110	$I4_1cd$	C_{4v}^{12}	$C_2(8)$
111	$P\bar{4}2m$	D_{2d}^1	$4D_{2d}$; $2D_2(2)$; $2C_{2v}(2)$; $5C_2(4)$; $C_s(4)$
112	$P\bar{4}2c$	D_{2d}^2	$4D_2(2)$; $2S_4(2)$; $7C_2(4)$
113	$P\bar{4}2_1m$	D_{2d}^3	$2S_4(2)$; $C_{2v}(2)$; $C_2(4)$; $C_s(4)$
114	$P\bar{4}2_1c$	D_{2d}^4	$2S_4(2)$; $2C_2(4)$
115	$P\bar{4}m2$	D_{2d}^5	$4D_{2d}$; $3C_{2v}(2)$; $2C_2(4)$; $2C_s(4)$
116	$P\bar{4}c2$	D_{2d}^6	$2D_2(2)$; $2S_4(2)$; $5C_2(4)$
117	$P\bar{4}b2$	D_{2d}^7	$2D_2(2)$; $2S_4(2)$; $4C_2(4)$
118	$P\bar{4}n2$	D_{2d}^8	$2D_2(2)$; $2S_4(2)$; $4C_2(4)$

Space group		Site symmetries
119 I$\bar{4}$m2	D_{2d}^{9}	4D_{2d}(2); 2C_{2v}(4); 2C_2(8); C_s(8)
120 I$\bar{4}$c2	D_{2d}^{10}	2D_2(4); 2S_4(4); 4C_2(8)
121 I$\bar{4}$2m	D_{2d}^{11}	2D_{2d}(2); D_2(4); S_4(4); C_{2v}(4); 3C_2(8); C_s(8)
122 I$\bar{4}$2d	D_{2d}^{12}	2S_4(4); 2C_2(8)
123 P4/mmm	D_{4h}^{1}	4D_{4h}; 2D_{2h}(2); 2C_{4v}(2); 7C_{2v}(4); 5C_s(8)
124 P4/mcc	D_{4h}^{2}	2D_4(2); 2C_{4h}(2); D_2(4); C_{2h}(4); 2C_4(4); 4C_2(8); C_s(8)
125 P4/nbm	D_{4h}^{3}	2D_4(2); 2D_{2d}(2); 2C_{2h}(4); C_4(4); C_{2v}(4); 4C_2(8); C_s(8)
126 P4/nnc	D_{4h}^{4}	2D_4(2); D_2(4); S_4(4); C_4(4); C_i(8); 4C_2(8)
127 P4/mbm	D_{4h}^{5}	2C_{4h}(2); 2D_{2h}(2); C_4(4); 3C_{2v}(4); 3C_s(8)
128 P4/mnc	D_{4h}^{6}	2C_{4h}(2); C_{2h}(4); D_2(4); C_4(4); 2C_2(8); C_s(8)
129 P4/nmm	D_{4h}^{7}	2D_{2d}(2); C_{4v}(2); 2C_{2h}(4); C_{2v}(4); 2C_2(8); 2C_s(8)
130 P4/ncc	D_{4h}^{8}	D_2(4); S_4(4); C_4(4); C_i(8); 2C_2(8)
131 P4$_2$/mmc	D_{4h}^{9}	4D_{2h}(2); 2D_{2d}(2); 7C_{2v}(4); C_2(8); 3C_s(8)
132 P4$_2$/mcm	D_{4h}^{10}	2D_{2h}(2); 2D_{2d}(2); D_2(4); C_{2h}(4); 4C_{2v}(4); 3C_2(8); 2C_s(8)
133 P4$_2$/nbc	D_{4h}^{11}	3D_2(4); S_4(4); C_i(8); 5C_2(8)
134 P4$_2$/nnm	D_{4h}^{12}	2D_{2d}(2); 2D_2(4); 2C_{2h}(4); C_{2v}(4); 5C_2(8); C_s(8)
135 P4$_2$/mbc	D_{4h}^{13}	2C_{2h}(4); S_4(4); D_2(4); 3C_2(8); C_s(8)
136 P4$_2$/mnm	D_{4h}^{14}	2D_{2h}(2); C_{2h}(4); S_4(4); 3C_{2v}(4); C_2(8); 2C_s(8)
137 P4$_2$/nmc	D_{4h}^{15}	2D_{2d}(2); 2C_{2v}(4); C_i(8); C_2(8); C_s(8)
138 P4$_2$/ncm	D_{4h}^{16}	D_2(4); S_4(4); 2C_{2h}(4); C_{2v}(4); 3C_2(8); C_s(8)
139 I4/mmm	D_{4h}^{17}	2D_{4h}(2); D_{2h}(4); D_{2d}(4); C_{4v}(4); C_{2h}(8); 4C_{2v}(8); C_2(16); 3C_s(16)
140 I4/mcm	D_{4h}^{18}	D_4(4); D_{2d}(4); C_{4h}(4); D_{2h}(4); C_{2h}(8); C_4(8); 2C_{2v}(8); 2C_2(16); 2C_s(16)
141 I4$_1$/amd	D_{4h}^{19}	2D_{2d}(4); 2C_{2h}(8); C_{2v}(8); 2C_2(16); C_s(16)
142 I4$_1$/acd	D_{4h}^{20}	S_4(8); D_2(8); C_i(16); 3C_2(16)
143 P3	C_{3}^{1}	3C_3
146 R3	C_{3}^{4}	C_3
147 P$\bar{3}$	S_{6}^{1}	2S_6; 2C_3(2); 2C_i(3)
148 R$\bar{3}$	S_{6}^{2}	2S_6; C_3(2); 2C_i(3)
149 P312	D_{3}^{1}	6D_3; 3C_3(2); 2C_2(3)
150 P321	D_{3}^{2}	2D_3; 2C_3(2); 2C_2(3)
151 P3$_1$12	D_{3}^{3}	2C_2(3)
152 P3$_1$21	D_{3}^{4}	2C_2(3)
153 P3$_2$12	D_{3}^{5}	2C_2(3)
154 P3$_2$21	D_{3}^{6}	2C_2(3)

Space group		Site symmetries
155 R32	D_3^7	$2D_3$; $C_3(2)$; $2C_2(3)$
156 P3m1	C_{3v}^1	$3C_{3v}$; $C_s(3)$
157 P31m	C_{3v}^2	C_{3v}; $C_3(2)$; $C_s(3)$
158 P3c1	C_{3v}^3	$3C_3(2)$
159 P31c	C_{3v}^4	$2C_3(2)$
160 R3m	C_{3v}^5	C_{3v}; $C_s(3)$
161 R3c	C_{3v}^6	$C_3(2)$
162 $P\bar{3}1m$	D_{3d}^1	$2D_{3d}$; $2D_3(2)$; $C_{3v}(2)$; $2C_{2h}(3)$; $C_3(4)$; $2C_2(6)$; $C_s(6)$
163 $P\bar{3}1c$	D_{3d}^2	$3D_3(2)$; $S_6(2)$; $2C_3(4)$; $C_i(6)$; $C_2(6)$
164 $P\bar{3}m1$	D_{3d}^3	$2D_{3d}$; $2C_{3v}(2)$; $2C_{2h}(3)$; $2C_2(6)$; $C_s(6)$
165 $P\bar{3}c1$	D_{3d}^4	$D_3(2)$; $S_6(2)$; $2C_3(4)$; $C_i(6)$; $C_2(6)$
166 $R\bar{3}m$	D_{3d}^5	$2D_{3d}$; $C_{3v}(2)$; $2C_{2h}(3)$; $2C_2(6)$; $C_s(6)$
167 $R\bar{3}c$	D_{3d}^6	$D_3(2)$; $S_6(2)$; $C_3(4)$; $C_i(6)$; $C_2(6)$
168 P6	C_6^1	C_6; $C_3(2)$; $C_2(3)$
171 $P6_2$	C_6^4	$2C_2(3)$
172 $P6_4$	C_6^5	$2C_2(3)$
173 $P6_3$	C_6^6	$2C_3(2)$
174 $P\bar{6}$	C_{3h}^1	$6C_{3h}$; $3C_3(2)$; $2C_s(3)$
175 P6/m	C_{6h}^1	$2C_{6h}$; $2C_{3h}(2)$; $C_6(2)$; $2C_{2h}(3)$; $C_3(4)$; $C_2(6)$; $2C_s(6)$
176 $P6_3/m$	C_{6h}^2	$3C_{3h}(2)$; $S_6(2)$; $2C_3(4)$; $C_i(6)$; $C_s(6)$
177 P622	D_6^1	$2D_6$; $2D_3(2)$; $C_6(2)$; $2D_2(3)$; $C_3(4)$; $5C_2(6)$
178 $P6_122$	D_6^2	$2C_2(6)$
179 $P6_522$	D_6^3	$2C_2(6)$
180 $P6_222$	D_6^4	$4D_2(3)$; $6C_2(6)$
181 $P6_422$	D_6^5	$4D_2(3)$; $6C_2(6)$
182 $P6_322$	D_6^6	$4D_3(2)$; $2C_3(4)$; $2C_2(6)$
183 P6mm	C_{6v}^1	C_{6v}; $C_{3v}(2)$; $C_{2v}(3)$; $2C_s(6)$
184 P6cc	C_{6v}^2	$C_6(2)$; $C_3(4)$; $C_2(6)$
185 $P6_3cm$	C_{6v}^3	$C_{3v}(2)$; $C_3(4)$; $C_s(6)$
186 $P6_3mc$	C_{6v}^4	$2C_{3v}(2)$; $C_s(6)$
187 $P\bar{6}m2$	D_{3h}^1	$6D_{3h}$; $3C_{3v}(2)$; $2C_{2v}(3)$; $3C_s(6)$
188 $P\bar{6}c2$	D_{3h}^2	$3D_3(2)$; $3C_{3h}(2)$; $3C_3(4)$; $C_2(6)$; $C_s(6)$
189 $P\bar{6}2m$	D_{3h}^3	$2D_{3h}$; $2C_{3h}(2)$; $C_{3v}(2)$; $2C_{2v}(3)$; $C_3(4)$; $3C_s(6)$
190 $P\bar{6}2c$	D_{3h}^4	$D_3(2)$; $3C_{3h}(2)$; $2C_3(4)$; $C_2(6)$; $C_s(6)$

Space group		Site symmetries
191 P6/mmm	$\mathbf{D}^1_{6h}$	$2\mathbf{D}_{6h}$; $2\mathbf{D}_{3h}(2)$; $\mathbf{C}_{6v}(2)$; $2\mathbf{D}_{2h}(3)$; $\mathbf{C}_{3v}(4)$; $5\mathbf{C}_{2v}(6)$; $4\mathbf{C}_{s}(12)$
192 P6/mcc	$\mathbf{D}^2_{6h}$	$\mathbf{D}_{6}(2)$; $\mathbf{C}_{6h}(2)$; $\mathbf{D}_{3}(4)$; $\mathbf{C}_{3h}(4)$; $\mathbf{C}_{6}(4)$; $\mathbf{D}_{2}(6)$; $\mathbf{C}_{2h}(6)$; $\mathbf{C}_{3}(8)$; $3\mathbf{C}_{2}(12)$; $\mathbf{C}_{s}(12)$
193 $P6_3$/mcm	$\mathbf{D}^3_{6h}$	$\mathbf{D}_{3h}(2)$; $\mathbf{D}_{3d}(2)$; $\mathbf{C}_{3h}(4)$; $\mathbf{D}_{3}(4)$; $\mathbf{C}_{3v}(4)$; $\mathbf{C}_{2h}(6)$; $\mathbf{C}_{2v}(6)$; $\mathbf{C}_{3}(8)$; $\mathbf{C}_{2}(12)$; $2\mathbf{C}_{s}(12)$
194 $P6_3$/mmc	$\mathbf{D}^4_{6h}$	$\mathbf{D}_{3d}(2)$; $3\mathbf{D}_{3h}(2)$; $2\mathbf{C}_{3v}(4)$; $\mathbf{C}_{2h}(6)$; $\mathbf{C}_{2v}(6)$; $\mathbf{C}_{2}(12)$; $2\mathbf{C}_{s}(12)$
195 P23	$\mathbf{T}^1$	$2\mathbf{T}$; $2\mathbf{D}_{2}(3)$; $\mathbf{C}_{3}(4)$; $4\mathbf{C}_{2}(6)$
196 F23	$\mathbf{T}^2$	$4\mathbf{T}(4)$; $\mathbf{C}_{3}(16)$; $2\mathbf{C}_{2}(24)$
197 I23	$\mathbf{T}^3$	$\mathbf{T}(2)$; $\mathbf{D}_{2}(6)$; $\mathbf{C}_{3}(8)$; $2\mathbf{C}_{2}(12)$
198 $P2_13$	$\mathbf{T}^4$	$\mathbf{C}_{3}(4)$
199 $I2_13$	$\mathbf{T}^5$	$\mathbf{C}_{3}(8)$; $\mathbf{C}_{2}(12)$
200 Pm3	$\mathbf{T}^1_{h}$	$2\mathbf{T}_{h}$; $2\mathbf{D}_{2h}(3)$; $4\mathbf{C}_{2v}(6)$; $\mathbf{C}_{3}(8)$; $2\mathbf{C}_{s}(12)$
201 Pn3	$\mathbf{T}^2_{h}$	$\mathbf{T}(2)$; $2\mathbf{S}_{6}(4)$; $\mathbf{D}_{2}(6)$; $\mathbf{C}_{3}(8)$; $2\mathbf{C}_{2}(12)$
202 Fm3	$\mathbf{T}^3_{h}$	$2\mathbf{T}_{h}(4)$; $\mathbf{T}(8)$; $\mathbf{C}_{2h}(24)$; $\mathbf{C}_{2v}(24)$; $\mathbf{C}_{3}(32)$; $\mathbf{C}_{2}(48)$; $\mathbf{C}_{s}(48)$
203 Fd3	$\mathbf{T}^4_{h}$	$2\mathbf{T}(8)$; $2\mathbf{S}_{6}(16)$; $\mathbf{C}_{3}(32)$; $\mathbf{C}_{2}(48)$
204 Im3	$\mathbf{T}^5_{h}$	$\mathbf{T}_{h}(2)$; $\mathbf{D}_{2h}(6)$; $\mathbf{S}_{6}(8)$; $2\mathbf{C}_{2v}(12)$; $\mathbf{C}_{3}(16)$; $\mathbf{C}_{s}(24)$
205 Pa3	$\mathbf{T}^6_{h}$	$2\mathbf{S}_{6}(4)$; $\mathbf{C}_{3}(8)$
206 Ia3	$\mathbf{T}^7_{h}$	$2\mathbf{S}_{6}(8)$; $\mathbf{C}_{3}(16)$; $\mathbf{C}_{2}(24)$
207 P432	$\mathbf{O}^1$	$2\mathbf{O}$; $2\mathbf{D}_{4}(3)$; $2\mathbf{C}_{4}(6)$; $\mathbf{C}_{3}(8)$; $3\mathbf{C}_{2}(12)$
208 $P4_232$	$\mathbf{O}^2$	$\mathbf{T}(2)$; $2\mathbf{D}_{3}(4)$; $3\mathbf{D}_{2}(6)$; $\mathbf{C}_{3}(8)$; $5\mathbf{C}_{2}(12)$
209 F432	$\mathbf{O}^3$	$2\mathbf{O}(4)$; $\mathbf{T}(8)$; $\mathbf{D}_{2}(24)$; $\mathbf{C}_{4}(24)$; $\mathbf{C}_{3}(32)$; $3\mathbf{C}_{2}(48)$
210 $F4_132$	$\mathbf{O}^4$	$2\mathbf{T}(8)$; $2\mathbf{D}_{3}(16)$; $\mathbf{C}_{3}(32)$; $2\mathbf{C}_{2}(48)$
211 I432	$\mathbf{O}^5$	$\mathbf{O}(2)$; $\mathbf{D}_{4}(6)$; $\mathbf{D}_{3}(8)$; $\mathbf{D}_{2}(12)$; $\mathbf{C}_{4}(12)$; $\mathbf{C}_{3}(16)$; $3\mathbf{C}_{2}(24)$
212 $P4_332$	$\mathbf{O}^6$	$2\mathbf{D}_{3}(4)$; $\mathbf{C}_{3}(8)$; $\mathbf{C}_{2}(12)$
213 $P4_132$	$\mathbf{O}^7$	$2\mathbf{D}_{3}(4)$; $\mathbf{C}_{3}(8)$; $\mathbf{C}_{2}(12)$
214 $I4_132$	$\mathbf{O}^8$	$2\mathbf{D}_{3}(8)$; $2\mathbf{D}_{2}(12)$; $\mathbf{C}_{3}(16)$; $3\mathbf{C}_{2}(24)$
215 $P\bar{4}3m$	$\mathbf{T}^1_{d}$	$2\mathbf{T}_{d}$; $2\mathbf{D}_{2d}(3)$; $\mathbf{C}_{3v}(4)$; $2\mathbf{C}_{2v}(6)$; $\mathbf{C}_{2}(12)$; $\mathbf{C}_{s}(12)$
216 $F\bar{4}3m$	$\mathbf{T}^2_{d}$	$4\mathbf{T}_{d}(4)$; $\mathbf{C}_{3v}(16)$; $2\mathbf{C}_{2v}(24)$; $\mathbf{C}_{s}(48)$
217 $I\bar{4}3m$	$\mathbf{T}^3_{d}$	$\mathbf{T}_{d}(2)$; $\mathbf{D}_{2d}(6)$; $\mathbf{C}_{3v}(8)$; $\mathbf{S}_{4}(12)$; $\mathbf{C}_{2v}(12)$; $\mathbf{C}_{2}(24)$; $\mathbf{C}_{s}(24)$
218 $P\bar{4}3n$	$\mathbf{T}^4_{d}$	$\mathbf{T}(2)$; $\mathbf{D}_{2}(6)$; $2\mathbf{S}_{4}(6)$; $\mathbf{C}_{3}(8)$; $3\mathbf{C}_{2}(12)$
219 $F\bar{4}3c$	$\mathbf{T}^5_{d}$	$2\mathbf{T}(8)$; $2\mathbf{S}_{4}(24)$; $\mathbf{C}_{3}(32)$; $2\mathbf{C}_{2}(48)$
220 $I\bar{4}3d$	$\mathbf{T}^6_{d}$	$2\mathbf{S}_{4}(12)$; $\mathbf{C}_{3}(16)$; $\mathbf{C}_{2}(24)$
221 Pm3m	$\mathbf{O}^1_{h}$	$2\mathbf{O}_{h}$; $2\mathbf{D}_{4h}(3)$; $2\mathbf{C}_{4v}(6)$; $\mathbf{C}_{3v}(8)$; $3\mathbf{C}_{2v}(12)$; $3\mathbf{C}_{s}(24)$
222 Pn3n	$\mathbf{O}^2_{h}$	$\mathbf{O}(2)$; $\mathbf{D}_{4}(6)$; $\mathbf{S}_{6}(8)$; $\mathbf{S}_{4}(12)$; $\mathbf{C}_{4}(12)$; $\mathbf{C}_{3}(16)$; $2\mathbf{C}_{2}(24)$
223 Pm3n	$\mathbf{O}^3_{h}$	$\mathbf{T}_{h}(2)$; $\mathbf{D}_{2h}(6)$; $2\mathbf{D}_{2d}(6)$; $\mathbf{D}_{3}(8)$; $3\mathbf{C}_{2v}(12)$; $\mathbf{C}_{3}(16)$; $\mathbf{C}_{2}(24)$; $\mathbf{C}_{s}(24)$
224 Pn3m	$\mathbf{O}^4_{h}$	$\mathbf{T}_{d}(2)$; $2\mathbf{D}_{3d}(4)$; $\mathbf{D}_{2d}(6)$; $\mathbf{C}_{3v}(8)$; $\mathbf{D}_{2}(12)$; $\mathbf{C}_{2v}(12)$; $3\mathbf{C}_{2}(24)$; $\mathbf{C}_{s}(24)$

Space group			Site symmetries
225	Fm3m	O_h^5	$2O_h(4)$; $T_d(8)$; $D_{2h}(24)$; $C_{4v}(24)$; $C_{3v}(32)$; $3C_{2v}(48)$; $2C_s(96)$
226	Fm3c	O_h^6	$O(8)$; $T_h(8)$; $D_{2d}(24)$; $C_{4h}(24)$; $C_{2v}(48)$; $C_4(48)$; $C_3(64)$; $C_2(96)$; $C_s(96)$
227	Fd3m	O_h^7	$2T_d(8)$; $2D_{3d}(16)$; $C_{3v}(32)$; $C_{2v}(48)$; $C_s(96)$; $C_2(96)$
228	Fd3c	O_h^8	$T(16)$; $D_3(32)$; $S_6(32)$; $S_4(48)$; $C_3(64)$; $2C_2(96)$
229	Im3m	O_h^9	$O_h(2)$; $D_{4h}(6)$; $D_{3d}(8)$; $D_{2d}(12)$; $C_{4v}(12)$; $C_{3v}(16)$; $2C_{2v}(24)$; $C_2(48)$; $2C_s(48)$
230	Ia3d	O_h^{10}	$S_6(16)$; $D_3(16)$; $D_2(24)$; $S_4(24)$; $C_3(32)$; $2C_2(48)$

8 Spectroscopic aspects of group theory

8.1 Selection rules and the transformations of operators

If a transition between two eigenstates, Ψ_1 and Ψ_2, connected by an operator P, is to have non-vanishing probability, at least one matrix element of the type $\langle\Psi_1|P|\Psi_2\rangle$ must differ from zero. The corresponding symmetry condition is that the triple direct product $\Gamma_{\Psi_1} \times \Gamma_P \times \Gamma_{\Psi_2^*}$ must contain the totally symmetric representation of the relevant point group. It follows that at least one component of Γ_P, the reduced representation of the operator P, must be contained in the direct product representation, $\Gamma_{\Psi_1} \times \Gamma_{\Psi_2}$, spanned by the two eigenstates. The rules for direct product formation are given in the next sections, §8.2 and §8.3. The transformation Γ_P for the more common types of operator may be obtained directly from the character tables as indicated below.

Operator, P	Γ_P transforms as	Character of Γ_P
Electric dipole moment (vector)	x, y, z	$\pm 1 + 2\cos\theta$
Electric quadrupole moment (symmetric 2nd rank tensor of zero trace)	x^2, y^2, z^2, xy, xz, yz (constraint:- $x^2 + y^2 + z^2 = 0$)	$2\cos\theta(\pm 1+2\cos\theta) - 1$
Electric polarisability (symmetric 2nd rank tensor)	x^2, y^2, z^2, xy, xz, yz	$2\cos\theta(\pm 1+2\cos\theta)$
First hyperpolarisability (3rd rank tensor)[8]	x^3, y^3, z^3, x^2y, xy^2, x^2z, xz^2, y^2z, yz^2, xyz	$2\cos\theta(4\cos^2\theta \pm 2\cos\theta - 1)$
Magnetic dipole moment (axial vector)	R_x, R_y, R_z	$1 \pm 2\cos\theta$

8.2 Rules for direct products

The characters of the representations $\Gamma_{j\times k}$, spanned by a direct product $\Gamma_j \times \Gamma_k$, are obtained by multiplying corresponding characters of the contributing representations:

$$\chi_{j\times k}(R) = \chi_j(R) . \chi_k(R) .$$

The representation is then reduced (if necessary) by the usual formula,

$$a_i = \frac{1}{h} \sum_R g_R . \chi_{j\times k}(R) . \chi_i(R) .$$

The rules of 'direct product algebra' are summarised below.

General rules

$A \times A = A$; $A \times B = B$; $A \times E_k = E_k$; $A \times F = F$; etc.

$B \times B = A$; $B \times E = E$; (regardless of other suffices).

Except in point groups D_2 and D_{2h} where $B \times B = B$ if the subscript numbers are not the same.

$g \times g = g$; $u \times u = g$; $g \times u = u$.

$' \times ' = '$; $'' \times '' = '$; $' \times '' = ''$.

Subscripts on A or B

$$1 \times 1 = 1; \quad 2 \times 2 = 1; \quad 1 \times 2 = 2.$$

Except in point groups $\mathbf{D_2}$ and $\mathbf{D_{2h}}$ where,

$$1 \times 2 = 3; \quad 2 \times 3 = 1; \quad 1 \times 3 = 2.$$

Tables for products between degenerate species

The point groups to which any one of the following tables applies are denoted by **n**, the order of the principal axis; this is taken to signify all the groups:

$$\mathbf{C_n}, \mathbf{C_{nh}}, \mathbf{C_{nv}}, \mathbf{D_n}, \mathbf{D_{nh}}, \mathbf{D_{nd}} \text{ and } \mathbf{S_n} \text{ for a given } \mathbf{n}.$$

Any exceptions or additions are explicitly stated. The relations hold regardless of any subscript on B. For groups in the set which have A, B, E or F without subscripts, put $A_1 = A_2 = A$ etc. The anti-symmetric component of the direct product (§8.3) is placed in parentheses.

3 and $\mathbf{S_6}$	E
E	$A_1 + (A_2) + E$

4 and $\mathbf{D_{2d}}$ not $\mathbf{D_{4d}}$	B	E
B	A	E
E		$A_1 + (A_2) + B_1 + B_2$

5 and $\mathbf{S_{10}}$	E_1	E_2
E_1	$A_1+(A_2)+E_2$	E_1+E_2
E_2		$A_1+(A_2)+E_1$

6 not $\mathbf{D_{6d}}$ not $\mathbf{S_6}$	B	E_1	E_2
B	A	E_2	E_1
E_1		$A_1+(A_2)+E_2$	$B_1+B_2+E_1$
E_2			$A_1+(A_2)+E_2$

7	E_1	E_2	E_3
E_1	$A_1+(A_2)+E_2$	E_1+E_3	E_2+E_3
E_2		$A_1+(A_2)+E_3$	E_1+E_2
E_3			$A_1+(A_2)+E_1$

8 and $\mathbf{D_{4d}}$ not $\mathbf{D_{8d}}$	B	E_1	E_2	E_3
B	A	E_3	E_2	E_1
E_1		$A_1+(A_2)+E_2$	E_1+E_3	$B_1+B_2+E_2$
E_2			$A_1+(A_2)+B_1+B_2$	E_1+E_3
E_3				$A_1+(A_2)+E_2$

D_{6d} and S_{12}	B	E_1	E_2	E_3	E_4	E_5
B	A	E_5	E_4	E_3	E_2	E_1
E_1		$A_1+(A_2)+E_2$	E_1+E_3	E_2+E_4	E_3+E_5	$B_1+B_2+E_4$
E_2			$A_1+(A_2)+E_4$	E_1+E_5	$B_1+B_2+E_2$	E_3+E_5
E_3				$A_1+(A_2)+B_1+B_2$	E_1+E_5	E_2+E_4
E_4					$A_1+(A_2)+E_4$	E_1+E_3
E_5						$A_1+(A_2)+E_2$

Cubic	E	F_1	F_2
E	$A_1+(A_2)+E$	F_1+F_2	F_1+F_2
F_1		$A_1+E+(F_1)+F_2$	$A_2+E+F_1+F_2$
F_2			$A_1+E+(F_1)+F_2$

Icosahedral	F_1	F_2	G	H
F_1	$A+(F_1)+H$	G+H	F_2+G+H	F_1+F_2+G+H
F_2		$A+(F_2)+H$	F_1+G+H	F_1+F_2+G+H
G			$A+(F_1+F_2)+G+H$	F_1+F_2+G+2H
H				$A+(F_1+F_2+G)+G+2H$

Linear $C_{\infty v}$, $D_{\infty h}$	Σ	Π	Δ
Σ	Σ	Π	Δ
Π		$\Sigma^++(\Sigma^-)+\Delta$	$\Pi+\Phi$
Δ			$\Sigma^++(\Sigma^-)+\Gamma$

(+.×.+ = +; +.×.- = -; -.×.- = +)

Direct products in the three dimensional rotation and rotation-inversion groups, **R(3)** and **O(3)** are given by the Clebsch-Gordan formula:

$$D^{(j_1)} \times D^{(j_2)} = D^{(j_1+j_2)} + D^{(j_1+j_2-1)} + \ldots + D^{|j_1-j_2|}$$

$$[D^{(j)} \times D^{(j)}]^+ = D^{(2j)} + D^{(2j-2)} + \ldots + D^{(0)} \text{ or } D^{(1)}$$

$$[D^{(j)} \times D^{(j)}]^- = D^{(2j-1)} + D^{(2j-3)} + \ldots + D^{(1)} \text{ or } D^{(0)}$$

8.3 Symmetric and anti-symmetric direct products

The direct product of a k-fold degenerate species, Γ_k, with itself (of dimension k^2) may be resolved (except in the trivial case of $k = 1$) into two components: a symmetric direct product, $[\Gamma_k^2]^+$, (of dimension $\frac{1}{2}k(k+1)$) and an anti-symmetric direct product, $[\Gamma_k^2]^-$, (of dimension $\frac{1}{2}k(k-1)$) thus:

$$\Gamma_k^2 = [\Gamma_k^2]^+ + [\Gamma_k^2]^- .$$

In vibrational spectroscopy, the symmetry species of the overtones of a degenerate fundamental are obtained from symmetric direct products, $[\Gamma_k^n]^+$. In the determination of electronic terms arising from the multiple occupation of equivalent one-electron orbitals in the strong crystal field approximation, the symmetric and anti-symmetric direct products for orbital angular momentum are taken with the appropriate spin functions to ensure that the total wave functions are anti-symmetric. The following formulae give the characters, $\chi^+_{\Gamma^2}(R)$ and $\chi^-_{\Gamma^2}(R)$, of the first symmetric and anti-symmetric direct products respectively of the species Γ^2 under an operation R. Higher symmetric direct products, Γ^n, for $\Gamma = E$ or F, may be obtained by means of the recursion formulae.

N. B. For ease of reference, the anti-symmetric components of the direct products tabulated in §8. 2 are placed in parentheses.

First overtones (n = 2) Γ = E, F, G, H, ... ($k \geq 2$)

$$\chi^{\pm}_{\Gamma^2}(R) = \tfrac{1}{2}\{[\chi_\Gamma(R)]^2 \pm \chi_\Gamma(R^2)\} .$$

For $\Gamma = E$ only, higher symmetric direct products are given by:

$$\chi^+_{E^n}(R) = \tfrac{1}{2}\{\chi_{E^{n-1}}(R).\,\chi_E(R) + \chi_E(R^n)\}$$

Second overtones (n = 3) Γ = F, G, H, ... ($k \geq 3$)

$$\chi^+_{\Gamma^3}(R) = \tfrac{1}{6}\{[\chi_\Gamma(R)]^3 + 3\chi_\Gamma(R).\,\chi_\Gamma(R^2) + 2\chi_\Gamma(R^3)\} .$$

For $\Gamma = F$ only, higher symmetric direct products are given by:

$$\chi^+_{F^n}(R) = \tfrac{1}{3}\{2\chi_{F^{n-1}}(R).\,\chi_F(R) - \tfrac{1}{2}([\chi_F(R)]^2 - \chi_F(R^2)).\,\chi_{F^{n-2}}(R) + \chi_F(R^n)\} .$$

For higher overtone levels of G and H species, see Tisza.[12]

8. 4 Conventions in spectroscopic notation

Summarised here in bold type are some of the recommendations[13] made by the Joint Commission for Spectroscopy of the International Astronomical Union and the International Union of Pure and Applied Physics, which deal with matters related to symmetry. Comments in ordinary type are our own and therefore carry no authority.

Species symbols. It is recommended that lower case letters be used for the species symbols of:

(a) normal vibrational modes,

and (b) one-electron orbital wave-functions.

In contrast, capital letters should be used for:

(c) vibrational state wave-functions,

(d) electronic state wave-functions,

and (e) vibronic state wave-functions.

The multiplicity of an electronic state (2S + 1) may be added as a left superscript, and the J value of a multiplet component as a right subscript, in parentheses if necessary.

The species symbols used are to be those set out by Herzberg.[14] Our tables conform with this recommendation, except in the case of point group $\mathbf{S_6}$ where we use A_g, A_u, E_g, E_u instead of Herzberg's A_g, B_u, E_{2g}, E_{1u}. As symbols for species of 3-, 4-, and 5-fold degeneracy we use F, G, and H, respectively, in accordance with the conventions of vibrational spectroscopy. However we note that the letters T, U, and V to denote these same species are clearly preferable when describing the electronic states or orbitals of atoms in crystal fields. Accordingly we have adopted the latter convention only in §8.6, there to avoid confusion with the customary symbols (S, P, D, F, G, H, ...) for atomic terms.

Choice of molecular axes. In the axial groups the z direction is always chosen as the principal axis of symmetry. In the cubic groups we have chosen it to coincide with one of the C_2 axes (in $\mathbf{T}$, $\mathbf{T_h}$, $\mathbf{T_d}$) or one of the C_4 axes (in $\mathbf{O}$, $\mathbf{O_h}$); in the icosahedral groups, with one of the C_5 axes. **The choice of x and y axes in some point groups remains arbitrary, and in the case of point groups C_{nv}, D_n, and D_{nh} (where n is even) the subscript number on the B species symbols is directly affected by the choice. The following recommendations, while not providing a general solution, do bring a conformity into the important class of planar molecules belonging to point groups C_{2v}, D_{2h}, D_{4h}, and D_{6h}.**

Point group C_{2v}: the x axis should be chosen perpendicular to the molecular plane.

Point group D_{2h}: the x axis should be chosen perpendicular to the molecular plane and the z axis so that it passes through the greatest number of atoms, or, if this rule is not decisive, so that it cuts the greatest number of bonds.

Point groups D_{4h} and D_{6h}: the C_2' axes shall pass through the greater numbers of atoms or, if this rule is not decisive, shall intersect the greater numbers of bonds. If it is necessary to define the x and y axes in these point groups, we choose the x axis to lie along one of the C_2' axes.

For non-planar molecules of point groups $\mathbf{C_{nv}}$ (n even), it is recommended that the σ_v planes be chosen to pass through the greater number of atoms, if this is not decisive, to intersect the greater number of bonds. If it is necessary to define the x and y axes in these point groups, we choose the x axis to lie in one of the σ_v planes.

Numbering of vibrational fundamentals. The symbol ν_k may denote both the kth normal mode and the frequency (in cm^{-1}) of its corresponding fundamental band. The sequence of numbering should begin with ν_1 for the totally symmetric mode of highest frequency, and continue with the modes in descending frequency within each symmetry species; the order of the species being that set out in Herzberg's tables.[14] It must be stressed that Herzberg's ordering differs from that adopted for the present tables (§6) in some cases. We have allowed the separation into g and u, or ' and " species to take precedence over the sequence A, B, E, F; thus showing explicitly the structure of the direct product groups.

If the motion in a particular normal mode is known to be largely localised (a 'group vibration') the following descriptive symbols may also be used in conjunction with the atomic symbols for the groups involved:

ν	**bond stretching,**
δ	**deformation or angle bending,**
τ	**twisting or torsion,**
ω	**wagging,**
ρ	**rocking.**

8.5 Elements of the Wilson G matrix

The kinetic energy matrix elements connecting two internal coordinates have been derived and tabulated by Wilson, Decius and Cross.[15] Their notation, which we follow here, uses the following conventions:

- μ_i denotes the reciprocal of the mass of the ith atom;
- ρ_{ij} denotes the reciprocal of the distance between atoms i and j;
- r denotes a bond stretching internal coordinate;
- ϕ denotes an angle deformation internal coordinate.

A kinetic energy matrix element carries a double subscript to indicate the types of internal coordinate involved, and a superscript giving the number of atoms common to the two coordinates. A further distinction is achieved by means of the following diagrammatic convention:

> Atoms common to both coordinates are indicated by double circles, and are always placed in a horizontal line; the non-common atoms are indicated along 45° diagonals (regardless of the actual geometry), those above the common set belong to the first coordinate, those below to the second.

A symbol, consisting of a pair of numbers in parentheses, may then be added to the **G** matrix element; the upper and lower numbers correspond to the number of non-common atoms along the upper and lower left diagonals respectively.

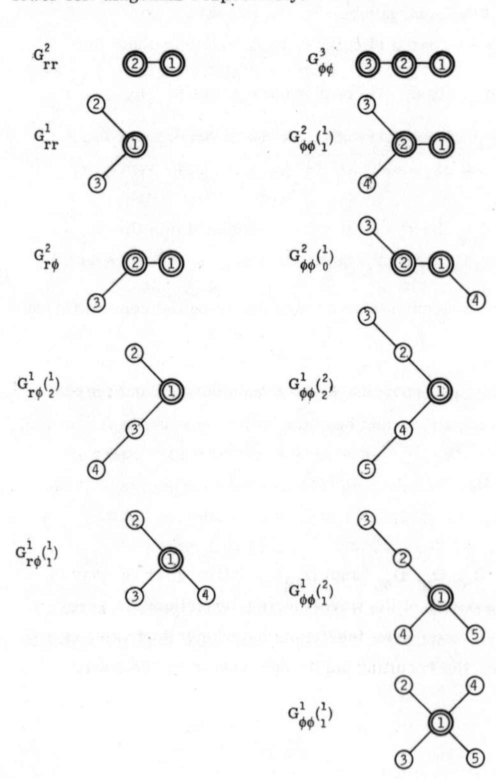

Fig. 3. Schematic representation of kinetic energy matrix elements. (Reproduced, by permission, from Molecular Vibrations by E. B. Wilson, J. C. Decius and P. C. Cross: McGraw-Hill, New York, 1955.)

$$\cos\psi_{\alpha\beta\gamma} = \frac{\cos\phi_{\alpha\delta\gamma} - \cos\phi_{\alpha\delta\beta}\cos\phi_{\beta\delta\gamma}}{\sin\phi_{\alpha\delta\beta}\sin\phi_{\beta\delta\gamma}}$$

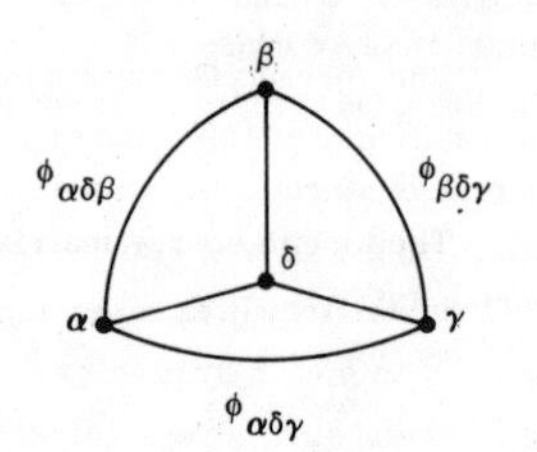

Fig. 4. Angles relating to the atomic positions α, β, γ, δ. (Source as fig. 3.)

$$\cos\psi_{\alpha\beta\gamma} = -\frac{\cos\phi_{\alpha\delta\gamma} - \cos\phi_{\alpha\delta\beta}\cos\phi_{\beta\delta\gamma}}{\sin\phi_{\alpha\delta\beta}\sin\phi_{\beta\delta\gamma}}.$$

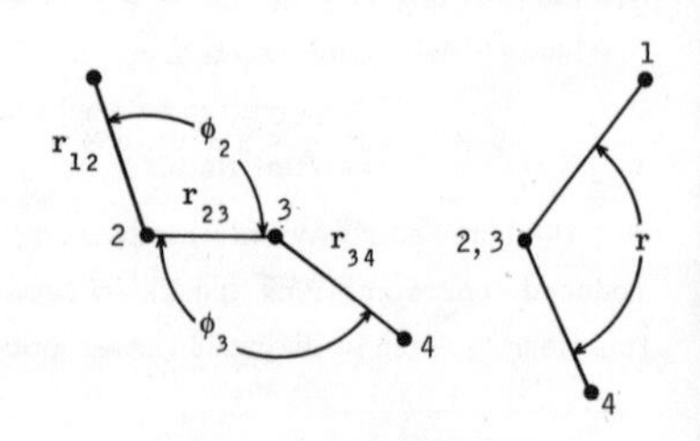

Fig. 5. Definition of the torsion angle, τ. (Source as fig. 3.)

General formulae for G_{rr}, $G_{r\phi}$, $G_{\phi\phi}$ elements (non-linear molecules)

G^2_{rr} $\quad \mu_1 + \mu_2$

G^1_{rr} $\quad \mu_1 \cos\phi$

$G^2_{r\phi}$ $\quad -\rho_{23}\mu_2 \sin\phi$

$G^1_{r\phi}\binom{1}{2}$ $\quad \rho_{13}\mu_1 \sin\phi_1 \cos\tau$

$\binom{1}{1}$ $\quad -(\rho_{13}\sin\phi_{213}\cos\psi_{234} + \rho_{14}\sin\phi_{214}\cos\psi_{243})\mu_1$

$G^3_{\phi\phi}$ $\quad \rho^2_{12}\mu_1 + \rho^2_{23}\mu_3 + (\rho^2_{12} + \rho^2_{23} - 2\rho_{12}\rho_{23}\cos\phi)\mu_2$

$G^2_{\phi\phi}\binom{1}{1}$ $\quad (\rho^2_{12}\cos\psi_{314})\mu_1 + [(\rho_{12} - \rho_{23}\cos\phi_{123} - \rho_{24}\cos\phi_{124})\rho_{12}\cos\psi_{314} +$

$\qquad (\sin\phi_{123}\sin\phi_{124}\sin^2\psi_{314} + \cos\phi_{324}\cos\psi_{314})\rho_{23}\rho_{24}]\mu_2$

$\binom{1}{0}$ $\quad -\rho_{12}\cos\tau[(\rho_{12} - \rho_{14}\cos\phi_1)\mu_1 + (\rho_{12} - \rho_{23}\cos\phi_2)\mu_2]$

$G^1_{\phi\phi}\binom{2}{2}$ $\quad -(\sin\tau_{25}\sin\tau_{34} + \cos\tau_{25}\cos\tau_{34}\cos\phi_1)\rho_{12}\rho_{14}\mu_1$

$\binom{2}{1}$ $\quad [(\sin\phi_{214}\cos\phi_{415}\cos\tau_{34} - \sin\phi_{215}\cos\tau_{35})\rho_{14} +$

$\qquad (\sin\phi_{215}\cos\phi_{415}\cos\tau_{35} - \sin\phi_{214}\cos\tau_{34})\rho_{15}]\dfrac{\rho_{12}\mu_1}{\sin\phi_{415}}$

$\binom{1}{1}$ $\quad [(\cos\phi_{415} - \cos\phi_{314}\cos\phi_{315} - \cos\phi_{214}\cos\phi_{215} + \cos\phi_{213}\cos\phi_{214}\cos\phi_{315})\rho_{12}\rho_{13} +$

$\qquad (\cos\phi_{413} - \cos\phi_{514}\cos\phi_{513} - \cos\phi_{214}\cos\phi_{213} + \cos\phi_{215}\cos\phi_{214}\cos\phi_{513})\rho_{12}\rho_{15} +$

$\qquad (\cos\phi_{215} - \cos\phi_{312}\cos\phi_{315} - \cos\phi_{412}\cos\phi_{415} + \cos\phi_{413}\cos\phi_{412}\cos\phi_{315})\rho_{14}\rho_{13} +$

$\qquad (\cos\phi_{213} - \cos\phi_{512}\cos\phi_{513} - \cos\phi_{412}\cos\phi_{413} + \cos\phi_{415}\cos\phi_{412}\cos\phi_{513})\rho_{14}\rho_{15}] \times$

$\qquad \dfrac{\mu_1}{\sin\phi_{214}\sin\phi_{315}}$

The angles ϕ, ψ and τ are defined in the accompanying figures.

The correct **G** matrix elements for linear molecules may be obtained from the general formula by substituting $\phi = 180°$ and $\tau = 0°$.[16]

The kinetic energy matrix elements for torsional coordinates have been derived and listed by Decius.[17]

8.6 Splitting of atomic terms in crystal fields

In the centro-symmetric point groups listed below (**I_h**, **O_h**, **D_{4h}** and **D_{3d}**) a suffix g or u may be added to the species symbols, as determined by the parity of the wave-function describing the free-atom term; this is expressed as a product of one-electron wave-functions, for which s, d, g, and i are **gerade** (g), and p, f, and h are **ungerade** (u), the resulting parity is governed by the usual produce rules:

$$g \times g = g; \quad g \times u = u; \quad u \times u = g .$$

For the non-centro-symmetric point groups (**T_d**, **D_{4d}** and **D_{2d}**) and the special case of **$D_{\infty h}$**, the reduced representations are listed separately for **gerade** (g) and **ungerade** (u) free-atom wave-functions. Term splittings in other point groups may be derived from those listed by means of the correlation tables in §7.2.

The table also serves to correlate the three-dimensional pure rotation, and rotation-inversion groups, **R(3)** and **O(3)** with point groups of lower symmetry.

Point group \ Term	$D^{(0)}$ S	$D^{(1)}$ P	$D^{(2)}$ D	$D^{(3)}$ F	$D^{(4)}$ G	$D^{(5)}$ H	$D^{(6)}$ I
I_h	A	T_1	V	T_2+U	$U+V$	T_1+T_2+V	$A+T_1+U+V$
O_h	A_1	T_1	$E+T_2$	$A_2+T_1+T_2$	$A_1+E+T_1+T_2$	$E+2T_1+T_2$	$A_1+A_2+E+T_1+2T_2$
T_d (g)	A_1	T_1	$E+T_2$	$A_2+T_1+T_2$	$A_1+E+T_1+T_2$	$E+2T_1+T_2$	$A_1+A_2+E+T_1+2T_2$
T_d (u)	A_2	T_2	$E+T_1$	$A_1+T_1+T_2$	$A_2+E+T_1+T_2$	$E+T_1+2T_2$	$A_1+A_2+E+2T_1+T_2$
D_{4h}	A_1	A_2+E	$A_1+B_1+B_2+E$	$A_2+B_1+B_2+2E$	$2A_1+A_2+B_1+B_2+2E$	$A_1+2A_2+B_1+B_2+3E$	$2A_1+A_2+2B_1+2B_2+3E$
D_{4d} (g)	A_1	A_2+E_3	$A_1+E_2+E_3$	$A_2+E_1+E_2+E_3$	$A_1+B_1+B_2+E_1+E_2+E_3$	$A_2+B_1+B_2+2E_1+E_2+E_3$	$A_1+B_1+B_2+2E_1+2E_2+E_3$
D_{4d} (u)	B_1	B_2+E_1	$B_1+E_1+E_2$	$B_2+E_1+E_2+E_3$	$A_1+A_2+B_1+E_1+E_2+E_3$	$A_1+A_2+B_2+E_1+E_2+2E_3$	$A_1+A_2+B_1+E_1+2E_2+2E_3$
D_{3d}	A_1	A_2+E	A_1+2E	A_1+2A_2+2E	$2A_1+A_2+3E$	A_1+2A_2+4E	$3A_1+2A_2+4E$
D_{2d} (g)	A_1	A_2+E	$A_1+B_1+B_2+E$	$A_2+B_1+B_2+2E$	$2A_1+A_2+B_1+B_2+2E$	$A_1+2A_2+B_1+B_2+3E$	$2A_1+A_2+2B_1+2B_2+3E$
D_{2d} (u)	B_1	B_2+E	$A_1+A_2+B_1+E$	$A_1+A_2+B_2+2E$	$A_1+A_2+2B_1+B_2+2E$	$A_1+A_2+B_1+2B_2+3E$	$2A_1+2A_2+2B_1+B_2+3E$
$D_{\infty h}$ (g)	Σ_g^+	$\Sigma_g^-+\Pi_g$	$\Sigma_g^++\Pi_g+\Delta_g$	$\Sigma_g^-+\Pi_g+\Delta_g+\Phi_g$	$\Sigma_g^++\Pi_g+\Delta_g+\Phi_g+\Gamma_g$	$\Sigma_g^-+\Pi_g+\Delta_g+\Phi_g+\Gamma_g+\Theta_g$	$\Sigma_g^++\Pi_g+\Delta_g+\Phi_g+\Gamma_g+\Theta_g+I_g$
$D_{\infty h}$ (u)	Σ_u^-	$\Sigma_u^++\Pi_u$	$\Sigma_u^-+\Pi_u+\Delta_u$	$\Sigma_u^++\Pi_u+\Delta_u+\Phi_u$	$\Sigma_u^-+\Pi_u+\Delta_u+\Phi_u+\Gamma_u$	$\Sigma_u^++\Pi_u+\Delta_u+\Phi_u+\Gamma_u+\Theta_u$	$\Sigma_u^-+\Pi_u+\Delta_u+\Phi_u+\Gamma_u+\Theta_u+I_u$

9 Group theory and molecular properties at equilibrium

9.1 Intrinsic physical properties of molecules and crystals

Intrinsic physical properties generally describe the relation between two quantities, each of which may be a scalar, a vector or a tensor. The number, n_A, of independent coefficients defining such a physical property is determined by the point group of the molecule or crystal, and is equal to the number of times that the totally symmetric representation appears in the reduced representation spanned by the physical property. When n_A is zero, a molecule or crystal in its equilibrium configuration (i. e. that of the point group assumed for it) cannot exhibit the property in question. Two methods have been used to find n_A: that of Bhagavantam and Suryanarayana[18] sets up the representation, $\Gamma_{\mathbf{T}}$, of the property from the transformation of the appropriate type of tensor, **T**, that defines it; the character, $\chi_{\mathbf{T}}(R)$, of the representation under an operation R (C_n^k or S_n^k) is given in the table below and may be reduced in any point group by means of the formula in §5. 2. The method due to Jahn[19] first finds the reduced representation of the property in the three-dimensional rotation-inversion group **O(3)** (see §6. 5); this may be obtained from direct products of the representation, $\Gamma_{\mathbf{v}}$, of a polar vector in **O(3)**, (see §8. 3) as indicated in the table. The representation in **O(3)** may then be reduced into any point group by means of correlation tables, and hence the number of times n_A that the totally symmetric representation appears may be found. (N. B. The correlations between the representations of **R(3)** or **O(3)** and those of other point groups are contained in §8. 6.) The table below lists only the minimum and maximum numbers of independent coefficients that the property may possess (the cases of perfect isotropy, **O(3)**, and complete anisotropy, $\mathbf{C}_1$, respectively) for a list of n_A values in each of the crystallographic point groups, and the properties of higher order tensors, see the works previously cited.[18-20] A pseudoscalar quantity (representation Γ_{ε}) has the transformation property of remaining invariant under a proper operation but of changing sign under an improper operation; i. e. the character, χ_{ε} (R) = ±1. The transformations of other 'pseudo' quantities follow directly. A pseudovector (representation $\Gamma_{\varepsilon} \times \Gamma_{\mathbf{v}}$) may also be called an axial vector; but is actually an anti-symmetric second rank tensor, $[\Gamma_{\mathbf{v}}^2]^-$.

Some useful generalisations can be made:

Enantiomorphism (or chirality) can only arise in molecules belonging to the symmetry point groups that contain no improper operations (S_n^k); i. e. the pure rotation groups:

$\mathbf{C}_n$, $\mathbf{D}_n$, **T**, **O** and **I**.

A permanent electric dipole moment can only arise where the molecular point group has a vector component (x, y or z) transforming as the totally symmetric representation. This behaviour is confined to the point groups whose symmetry elements do not meet at a single point, but in a line or plane; i. e. the groups $\mathbf{C}_n$ and $\mathbf{C}_{nv}$ (including $\mathbf{C}_{1v} = \mathbf{C}_s$).

Electric polarisability: the trace, $\alpha_{xx} + \alpha_{yy} + \alpha_{zz}$, of this symmetric second rank tensor always transforms as the totally symmetric representation of a point group; accordingly the polarisability can always be non-zero.

The first electric hyperpolarisability transforms as cubic functions of the vector components x, y and z, and is therefore always zero for systems having inversion symmetry (i).

Physical property represents relation between	Nature of physical property	Γ_T	$\chi_T(R)$ $R=C_n^k(+)$ or $S_n^k(-)$	Γ_T in O(3)	n_A in O(3)	n_A in C_1	Examples of physical property
Scalar and scalar	Scalar	1	+1	$D_g^{(0)}$	1	1	Density
Scalar and pseudoscalar	Pseudoscalar	Γ_ε	±1	$D_u^{(0)}$	0	1	Enantiomorphism
Scalar and vector	Vector	Γ_v	2c±1	$D_u^{(1)}$	0	3	Electric dipole moment; pyro-electricity
Pseudoscalar and vector	Pseudovector	$\Gamma_\varepsilon \times \Gamma_v = [\Gamma_v^2]^-$	1±2c	$D_g^{(1)}$	0	3	Magnetic dipole moment
Vector and vector; Scalar and symmetric 2nd rank tensor; Pseudovector and pseudovector	Symmetric 2nd rank tensor	$[\Gamma_v^2]^+$	2c(2c±1)	$D_g^{(0)}+D_g^{(2)}$	1	6	Electric polarisa-bility; thermal and electrical conduc-tivity; thermoelec-tricity. Thermal expansion. Magnetic susceptibility.
Vector and vector		-	2c(2c±1)-1	$D_g^{(2)}$	0	5	Electric quadru-pole moment.
Pseudoscalar and symmetric 2nd rank tensor	Symmetric 2nd rank pseudotensor	$\Gamma_\varepsilon \times [\Gamma_v^2]^+$	2c(1±2c)	$D_u^{(0)}+D_u^{(2)}$	0	6	Optical activity (gyration tensor)
Vector and square of vector	Symmetric 3rd rank tensor	$[\Gamma_v^3]^+$	$2c(4c^2±2c-1)$	$D_u^{(1)}+D_u^{(3)}$	0	10	First electric hyper-polarisability
Vector and symmetric 2nd rank tensor	3rd rank tensor	$\Gamma_v \times [\Gamma_v^2]^+$	$2c(2c±1)^2$	$2D_u^{(1)}+D_u^{(2)}+D_u^{(3)}$	0	18	Piezoelectricity; electro-optical Kerr effect
Symmetric 2nd rank tensor and symmetric 2nd rank tensor	Symmetric 4th rank tensor	$[([\Gamma_v^2]^+)^2]^+$	$16c^4±8c^3-4c^2+1$	$2D_g^{(0)}+2D_g^{(2)}+D_g^{(4)}$	2	21	Elasticity

$c = \cos 2\pi k/n$

The electric quadrupole moment will only be identically zero for molecules of the cubic and icosahedral point groups.

Rotational properties: the inertia tensor is symmetric and of second rank; its diagonal elements, I_{xx}, I_{yy} and I_{zz} are the moments of inertia and the off-diagonal elements, $-I_{xy}$, $-I_{xz}$, $-I_{yz}$, are the products of inertia. The matrix may be diagonalised by an orthogonal transformation (corresponding to a choice of axes coinciding with certain symmetry elements) to give the three **principal moments of inertia,** $I_a \le I_b \le I_c$.

Molecules are classified as types of rotor, according to the following table:

Type of rotor	Principal moments	Point groups
spherical top	$I_a = I_b = I_c$	$\mathbf{T}$, $\mathbf{T_h}$, $\mathbf{T_d}$, $\mathbf{O}$, $\mathbf{O_h}$, $\mathbf{I}$, $\mathbf{I_h}$.
symmetric top	$\left\{\begin{array}{l}\text{prolate: } I_a < I_b = I_c \\ \text{oblate: } I_a = I_b < I_c\end{array}\right\}$	$\mathbf{D_{2d}}$, $\mathbf{C_n}$, $\mathbf{C_{nv}}$, $\mathbf{C_{nh}}$, $\mathbf{D_n}$, $\mathbf{D_{nd}}$, $\mathbf{D_{nh}}$, $(n\geq3)$, $\mathbf{S_{2n}}$, $(n\geq2)$
asymmetric top	$I_a < I_b < I_c$	$\mathbf{C_1}$, $\mathbf{C_s}$, $\mathbf{C_i}$, $\mathbf{C_2}$, $\mathbf{C_{2v}}$, $\mathbf{C_{2h}}$, $\mathbf{D_2}$, $\mathbf{D_{2h}}$
linear	$I_a = 0$; $I_b = I_c$	$\mathbf{C_{\infty v}}$, $\mathbf{D_{\infty h}}$.

[N. B. Symmetry does not take account of the possibility of accidental equalities among I_a, I_b and I_c.]

The **symmetry number** of a molecule is the number of distinct, physically feasible ways of orienting it to give an indistinguishable configuration. For non-linear molecules, the symmetry number is equal to the order of the pure rotation subgroup. For linear molecules it is 2 ($\mathbf{D_{\infty h}}$) or 1 ($\mathbf{C_{\infty v}}$).

The **independent structural parameters** of a molecule belonging to a given point group are those interatomic distances and angles which must be fixed in order to completely define the geometry. Although the choice of independent parameters is not unique, their number, n_P, is: n_P is the number of internal degrees of freedom that preserve the full point group symmetry; i. e. n_P is equal to the number of totally symmetric normal modes of vibration of the molecule. This number may be found by the usual method (§5. 3) or, more rapidly, by applying the following rules.[21] The molecule is viewed as consisting of sets of symmetrically equivalent atoms (i. e. the members of such a set are permuted among themselves by the operations of the point group $\mathbf{H}$ [order h]). The local symmetry point group, $\mathbf{G_i}$, of a typical member of each set is found by inspection ($\mathbf{G_i}$ [order g_i] is either a subgroup, $\mathbf{C_n}$ or $\mathbf{C_{nv}}$, of $\mathbf{H}$, or is $\mathbf{H}$ itself if the set contains only one atom. The number, n_i, of atoms in the ith set is $n_i = h/g_i$). The number, P_i, of vector components x, y or z transforming as the totally symmetric representation of $\mathbf{G_i}$ is noted:

$\mathbf{G_i}$	P_i
$\mathbf{C_1}$	3
$\mathbf{C_s}$	2
$\mathbf{C_n}$ or $\mathbf{C_{nv}}$	1
all other point groups	0

The values of P_i are summed over all the subgroups $\mathbf{G_i}$; if $\mathbf{H}$ itself is $\mathbf{C_1}$, $\mathbf{C_s}$, $\mathbf{C_n}$ or $\mathbf{C_{nv}}$ the appropriate value of P_i is subtracted. The resulting number is n_P, the number of independent structural parameters.

9.2 Stereoisomerism and Polya's theorem

Analytical methods for the enumeration of isomers have been developed by Lunn and Senior[22] and by Polya,[23] using permutation group theory. Kennedy, McQuarrie and Brubaker[24] have reformulated Polya's method for stereoisomerism in terms of point groups; their method is summarised here.

The 'skeletal' parent structure of the molecule (having its atom sites unoccupied) belongs to point group **G**, of order g. The symmetry operations of **G** cause permutations within the set of point sites (the point sites need not necessarily be all symmetrically equivalent); a subset of points that are permuted only among themselves by a symmetry operation constitutes a **cycle**, and the number of such poings in the subset is the **length**, k, of the cycle. In general, a symmetry operation acting on a set of points will produce several cycles, some of differing length: the **number of cycles of length** k is denoted by j_k. A symmetry operation is then defined, in terms of its effect in permuting p points, by the collection of j_k values. The number of symmetry operations of **G** that consist of j_1 cycles of length 1, j_2 cycles of length 2, ... j_k cycles of length k, ... j_p cycles of length p, is written $h_{j_1 j_2 \ldots j_p}$. The **cycle index**, Z(G) of the group **G** applied to p points is defined by

$$Z(G) = \frac{1}{g} \sum h_{j_1 j_2 \ldots j_p} f_1^{j_1} f_2^{j_2} \ldots f_p^{j_p}$$

where $f_1, f_2 \ldots f_p$ are p variables, and the summation is taken over all the symmetry operations of **G**

Let $F_{n_1 n_2 \ldots n_m}$ be the number of stereoisomers arising when n_i substituents of type i occupy the sites of the parent skeleton; m is the number of different substituents. The numbers $F_{n_1 n_2 \ldots n_m}$ are used to define a configuration counting series $F(x_1, x_2, \ldots x_m)$:

$$F(x_1, x_2, \ldots x_m) = \sum F_{n_1 n_2 \ldots n_m} x_1^{n_1} x_2^{n_2} \ldots x_m^{n_m}$$

where the summation is over $n_i = 0$ to ∞ for each i from 1 to m. $x_1, x_2, \ldots x_m$ are m variables. Polya's theorem states that

$$F(x_1, x_2, \ldots x_m) = Z[\mathbf{G}; f(x_1, x_2, \ldots x_m)]$$

where $Z[\mathbf{G}; f(x_1, x_2, \ldots x_m)]$ denotes the polynomial obtained from the cycle index Z(G) by replacing f_k with $(x_1^k + x_2^k + \ldots + x_m^k)$. Hence the numbers $F_{n_1 n_2 \ldots n_m}$ are found as the coefficients in the polynomial expansion of the cycle index.

This method gives the number of **geometrical** stereoisomers; the total number of stereoisomers (which distinguishes between enantiomorphs) is obtained by using the pure rotation subgroup for **G**.

The binomial expansion:

$$(x_1+x_2)^q = \binom{q}{0}x_1^q + \binom{q}{1}x_1^{q-1}x_2 + \binom{q}{2}x_1^{q-2}x_2^2 + \ldots + \binom{q}{q}x_2^q$$

$$\binom{q}{r} = \frac{q!}{r!\,(q-r)!}$$

The multinomial expansion:

$$(x_1 + x_2 + \ldots + x_m)^q = \sum \frac{q!}{q_1!\,q_2!\ldots q_m!} x_1^{q_1} x_2^{q_2} \ldots x_m^{q_m} .$$

The sum is over all non-negative integers q_i which satisfy $\sum_1^m q_i = q$.

9.3 The relative abundances of isotopically substituted species

If p symmetrically equivalent sites in a molecule are occupied by atoms of an element having two isotopes, E and E', the probability, A_x, of the molecular species containing x atoms of E and (p - x) atoms of E' is given by the binomial distribution

$$A_x = \binom{p}{x} r^x (1 - r)^{p-x}$$

where r and (1 - r) are the fractions of E and E', respectively, in the equilibrium composition of the bulk element.

N. B. This relation assumes that any isotope effect may be neglected. The natural abundances of isotopes are listed in §1. 3.

10 Bibliography and references

Bibliography

C. J. Ballhausen, Introduction to Ligand Field Theory. McGraw-Hill, New York (1962).

E. Bauer and P. H. E. Meijer, Group Theory, The Application to Quantum Mechanics. North-Holland Pub. Co. Amsterdam (1962).

S. Bhagavantam and T. Venkatarayudu, Theory of Groups and its Application to Physical Problems. Waltair, India (1948).

M. J. Buerger, X-Ray Crystallography. Wiley and Sons, New York (1942).

C. W. Bunn, Chemical Crystallography. Oxford University Press (1961).

F. A. Cotton, Chemical Applications of Group Theory. Interscience (1963).

A. P. Cracknell, Applied Group Theory. Pergamon Press, Oxford (1968).

H. Eyring, J. Walter and G. E. Kimball, Quantum Chemistry. Wiley and Sons, New York (1944).

L. M. Falicov, Group Theory and its Physical Applications. Chicago University Press (1966).

U. Fano and G. Racah, Irreducible Tensorial Sets. Academic Press, New York (1959).

J. S. Griffith, The Irreducible Tensor Method for Molecular Symmetry Groups. Prentice-Hall, London (1962).

G. G. Hall, Applied Group Theory. Longmans, London (1967).

M. Hamermesh, Group Theory and its application to Physical Problems. Addison-Wesley, London (1962).

V. Heine, Group Theory in Quantum Mechanics. Pergamon Press, Oxford (1960).

G. Herzberg, Infrared and Raman Spectra of Polyatomic Molecules. Van Nostrand, New York (1945).

G. Herzberg, Electronic Spectra of Polyatomic Molecules. Van Nostrand Reinhold Co., New York (1966).

B. Higman, Applied Group-Theoretic and Matrix Methods. Dover, New York (1964).

H. H. Jaffé and M. Orchin, Symmetry in Chemistry. Wiley and Sons, New York (1965).

J. W. Leech and D. J. Newman, How to use Groups. Methuen, London (1969).

J. S. Lomont, Applications of Finite Groups. Academic Press, New York (1959).

R. McWeeny, Symmetry, an introduction to Group Theory and its Applications. Pergamon Press, Oxford (1963).

M. E. Rose, Elementary Theory of Angular Momentum. John Wiley and Son, New York (1957).

D. S. Schonland, Molecular Symmetry, an introduction to Group Theory and its uses in Chemistry. Van Nostrand, New York (1965).

M. Tinkham, Group Theory and Quantum Mechanics. McGraw-Hill, New York (1964).

H. Watanabe, Operator Methods in Ligand Field Theory. Prentice-Hall, New Jersey (1966).

E. P. Wigner, Group Theory. Academic Press Inc., New York (1959).

E. B. Wilson, Jr., J. C. Decius and P. C. Cross, Molecular Vibrations. McGraw-Hill, New York (1955).

References

1. B. N. Taylor, W. H. Parker and D. N. Langenberg, Rev. Mod. Phys. **41**, 477 (1969).
2. Symbols, Signs and Abbreviations. Royal Society, London (1969). Pure and Appl. Chem. **9**, 453 (1964).

3. Pure and Appl. Chem. **20**, 4 (1970).
25th Compt. Rend. IUPAC (1970).
4. N. N. Greenwood, Chem. in Brit. **6**, 119 (1970).
5. H. M. Cundy and A. P. Rollett, Mathematical Models. Oxford University Press (1961).
6. S. M. Ferigle and A. G. Meister, Amer. J. Phys. **20**, 421 (1952).
7. J. E. Rosenthal and G. M. Murphy, Rev. Mod. Phys. **8**, 317 (1936).
8. S. J. Cyvin, J. E. Rauch and J. C. Decius, J. Chem. Phys. **43**, 4083 (1965).
9. S. F. A. Kettle and A. J. Smith, J. Chem. Soc. A 688 (1967).
10. R. S. Halford, J. Chem. Phys. **14**, 8 (1946).
H. Winston and R. S. Halford, J. Chem. Phys. **17**, 607 (1949).
11. International Tables for X-Ray Crystallography, vol. 1, K. Lonsdale (ed.). Kynoch Press, Birmingham (1959).
12. L. Tisza, Z. Physik, **82**, 48 (1933).
13. J. Chem. Phys. **23**, 1997 (1955).
14. G. Herzberg, Infrared and Raman Spectra of Polyatomic Molecules. Van Nostrand, New York (1945).
15. E. B. Wilson Jr., J. C. Decius and P. C. Cross, Molecular Vibrations. McGraw-Hill, New York (1955).
16. S. M. Ferigle and A. G. Meister, J. Chem. Phys. **19**, 982 (1951).
17. J. C. Decius, J. Chem. Phys. **16**, 1025 (1948).
18. S. Bhagavantam and D. Suryanarayana, Acta Cryst. **2**, 21 (1949).
19. H. A. Jahn, Z. Krystallogr. **98**, 191 (1937).
H. A. Jahn, Acta Cryst. **2**, 30 (1949).
20. B. Higman, Applied Group-Theoretic and Matrix Methods. Dover, New York (1964).
21. M. J. Ware and J. A. Salthouse, unpublished work.
22. A. C. Lunn and J. K. Senior, J. Phys. Chem. **33**, 1027 (1929).
23. G. Polya, Krist. **93**, 415 (1936).
24. B. A. Kennedy, D. A. McQuarrie and C. H. Brubaker Jr. Inorg. Chem. **3**, 265 (1964).